DESTINATION DISASTER

DESTINATION DISASTER

Aviation Accidents in the modern age

ANDREW BROOKES

Ian Allan

60th ANNIVERSARY

For Jake, his book

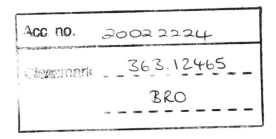

First published 2002

ISBN 0 7110 2862 1

Published by Ian Allan Publishing
an imprint of Ian Allan Publishing Ltd, Hersham, Surrey KT12 4RG.

Printed by Ian Allan Printing Ltd, Hersham, Surrey KT12 4RG.

Code: 0210/C

Front cover:
The tail of TWA800 standing all-forlorn in the Atlantic off East Moriches. *Associated Press*

Contents

Abbreviations and Glossary

AAIB	Air Accidents Investigation Branch
ACARS	Aircraft Communication Addressing & Reporting System
agl	above ground level
Airprox	Report submitted after suspected loss of safe aircraft separation
Airway	Designated air route, usually defined by ground-based radio beacons
ARFF	Aircraft Rescue & Fire Fighting
ARTCC	Air Route Traffic Control Centre
ASI	Airspeed Indicator
ATC	Air Traffic Control
AFCS	Automatic Flight Control & Augmentation System
ATS	Auto-throttle System
CVR	Cockpit Voice Recorder. Tape recorder that transcribes comments and audible actions of the flight crew
Decision Height	Specified altitude at which the crew must decide either to continue or abandon a landing approach
Decision Speed (V1)	Speed at which the crew must decide to continue or abandon a take-off
Decision Speed (V2)	Minimum speed required in the air after an engine failure at V1
DME	Distance measuring equipment. Measures in nautical miles an aircraft's slant range from the beacon
FAA	Federal Aviation Administration
FAS	Flight Augmentation System
FDR	Flight Data Recorder
Feather	Adjustment of an aircraft's propeller to reduce drag following engine stoppage
FL	Flight Level. A level of constant atmospheric pressure related to a specific pressure datum of 1013.2mb. Each level is stated in three digits, eg FL250 represents a barometric indication of 25,000ft; FL255 an indication of 25,500ft
FMS	Flight Management System
fpm	feet per minute
GCA	Ground Controlled Approach
GPWS	Ground Proximity Warning System
Hz	hertz
IAS	Indicated Air Speed
ICAO	International Civil Aviation Organisation
IFR	Instrument Flight Rules
ILS	Instrument Landing System. Standard landing aid comprising a glide slope beam for vertical and a localiser beam for lateral guidance
IMC	Instrument meteorological conditions; ie flying on instruments

LATCC	London Area Terminal Control Centre
MDA	Minimum Descent Altitude. The lowest altitude, expressed in feet above mean sea level, to which descent is authorised on final approach during a standard instrument procedure where no electronic glide slope is provided
mHz	megahertz
msl	mean sea level
NASA	National Aeronautics & Space Administration
nm	nautical mile. 1nm is about 1.15 land miles. 1kt is 1nm/hr
NTSB	National Transportation Safety Board
Octa	Unit of measurement of cloud cover, expressed in eighths
PAR	Precision Approach Radar
psi	lb/sq in
QFE	Atmospheric pressure at airfield datum
QNH	Altimeter setting to obtain elevation when on the ground
Rotate Speed (VR)	Speed for raising aircraft nose gear off the ground during take off
RVR	Runway Visual Range. Horizontal distance visible when looking down a runway centreline
SID	Standard Instrument Departure
Squawk	Transponder transmission to identify an aircraft
SSR	Secondary Surveillance Radar — see Transponder
STCA	Short Term Conflict Alert
TACAN	Tactical Air Navigation. Ground-based UHF aid which provides suitably equipped aircraft with continuous bearing and distance information
TAF	Terminal Aerodrome Forecast
TAS	True Air Speed
TCAS	Traffic Alert & Collision Avoidance System
Time	Reported in this book by means of the 24hr clock. So 16.16hrs is 4.16pm, and 16.16:23 is 23 seconds past 4.16pm
Transponder	Enables each suitably equipped aircraft to superimpose its individual designator on the radar return painted on a ground controller's screen. In enabling controllers to differentiate between 'blips' with accuracy, it confers a higher degree of safety in a crowded and high-speed environment such as an airway. If an emergency occurs, the pilot 'squawks' an Emergency Code to alert the ground agency
VFR	Visual Flight Rules. The pilot maintains aircraft separation by sight, usually independent of an ATC facility
VMC	Visual meteorological conditions
VOR	Very high frequency Omnidirectional Radio range
Wind shear	Currents representing a significant change, in terms of direction or speed, from the ground airflow

Introduction

On any given day, more than four million people around the world take to the air on one of the 38,000 flights the airlines operate. As you open this book, a quarter of a million people will be in flight.

The reassuring news is that the fatal accident rate per million flights for large passenger aircraft operations is now about half of what it was 10 years ago. In 1991 there were 1.7 fatal accidents per million flights, and in 1999 the rate dropped below one fatal accident per million flights for the first time. This reduced to 0.85 in 2000.

Statistics are funny things — the number of people seriously injured in the UK from over hastily putting on trousers, tights or socks exceeds those ascribed to chain saws by a factor of 10. So just let me say from the outset that flying is safer now than it has ever been. It is 20 times safer to get airborne in a commercial airliner than to drive to the airport in the first place. More people die from falling downstairs than from air crashes. Consequently, this is not a 'shock, horror' book. Rather, it tries to outline the background behind the most dramatic and note-worthy of recent flying accidents, to show the causes and to say what is being done to stop such accidents from happening again. I have flown enough aircraft types to know that flight safety is rarely straightforward, and in dealing with past accidents I have tried to pass on food for thought rather than pious platitudes.

Each nation has its independent aircraft accident investigation organisation, charged as in the case of the British Air Accidents Investigation Branch (AAIB) with 'determining the circumstances and causes of an accident with a view to the preservation of life and the avoidance of accidents in the future'. Like the American National Transportation Safety Board (NTSB), which investigates accidents befalling US-registered aircraft, the AAIB then makes recommendations to the aircraft operators for action, or to the national regulatory authority for permanent regulations, to lessen the probability of repetition. *Destination Disaster* could not have been written without access to the detailed investigations and conclusions contained within British, French and US accident reports. I am eternally grateful to the men and women who painstakingly compiled them, and if any false conclusions are drawn from their labours, the fault is mine alone.

Finally, as it is easy to be clever with hindsight, I hope that I do not give offence to aviators involved in accidents, or to their friends and relatives. I regard all those who had the misfortune to be involved in any flying accidents as colleagues who would want successive generations to learn from their experiences. *Destination Disaster* is certainly not written to point the finger at anyone — I have enough pilot hours under my belt to admit that, sometimes, there but for the Grace of God went I. If this book stops anyone from thinking that aircraft cannot bite, and stimulates thought as to what to do should that awful moment occur, it will have done its job.

Andrew Brookes
July 2002

1

'Concorde . . . You Have Strong Flames Behind You'

There has been only one successful supersonic airliner — the Anglo-French Concorde. Discussions began between BAC and Sud-Aviation about the specification for this supersonic transport in June 1961. By May 1964 the preliminary design of Concorde had been agreed. The French prototype took to the air in March 1969, followed by the first British Concorde a month later.

Concorde achieved Mach 2 (1,320mph) in late 1970 and made its first transatlantic flight in early September 1971. Full airworthiness certification came on 5 December 1975, and Concorde began airline services on 21 June 1976. It was to fly those with the money to pay to get somewhere fast without hindrance for the next 25 years.

Flying Concorde has always been special. Powered by an uprated version of the Olympus jet engines that powered the Vulcan strategic bomber, Concorde's delta-winged shape still makes people stop and stare. Inside the cabin, which is really only a silver tube to slide through the sound barrier, statesmen, billionaires, rock stars, models and other beautiful people sit cheek by jowl, as champagne is served, and share conversation because there is no in-flight video system. In spite of the narrow cabin, the supersonic jet is the preferred mode of transatlantic transport among the world's glitterati. Businessmen in a hurry regularly rub shoulders with Michael Jackson and Madonna, and Richard Gere was not untypical when he and his baby returned to New York on Concorde after a Paris holiday in spring 2000. An Air France Concorde crewmember waxed lyrical on behalf of many. 'Up there you are between the Moon and the Earth. You can see the curvature of the Earth, you are already in space. Up there you are constantly on the edge, never sure you will reach the other end.'

But it was not just the rich and famous who flew in Concorde. Taking groups of affluent tourists beyond the speed of sound in soft leather seats had become a major money-spinner, and on Tuesday 25 July 2000 Air France Concorde F-BTSC stood on the ramp at Paris Charles de Gaulle ready to take charter flight AF4590 to New York. In addition to the nine flight and cabin crew, 100 mostly German passengers were booked on the flight of a lifetime prior to joining a Caribbean cruise.

The Air France Concorde, known as Tango Sierra Charlie after the last three letters of its registration, had undergone a regular repair programme. Although it

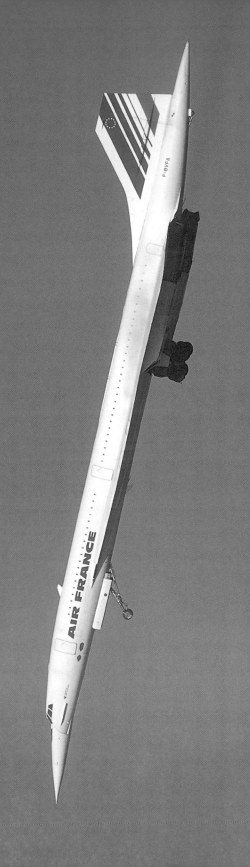

first flew on 31 January 1975, it was still relatively young, having clocked up only 11,989 flying hours. Its four Olympus 593 Mk610 jet engines were each rated at 38,000lb thrust, with reheat giving extra thrust at take-off. The engines had been removed the previous year and taken to Rolls-Royce in Cardiff, where they were stripped for a full examination in September. A shorter service took place on 28 April 2000. Although the French engineering union was prone to grumble over a shortage of ground engineers, Concorde maintenance crews were the elite. The team of 100 or so took pride in their work and looked after Concorde like a vintage Bugatti.

If any of the German passengers booked on the charter flight had concerns about the safety of supersonic flight, they would have been reassured by the commander of Tango Sierra Charlie that fateful day, Christian Marty. Tall, with brown hair, brown eyes and a ready smile, Marty was the first person to windsurf across the Atlantic. Throughout the 37-day crossing, he insisted on sleeping tied to his board despite being accompanied by a support boat. With Gallic panache, Capt Marty liked to describe himself as *'un grand contemplatif'* — a philosophical type — and boasted that 'there are fewer Concorde pilots than astronauts'.

After joining Air France in 1968, he had flown Boeing 727s and 737s, and Airbuses, and become an instructor, before joining the airline's roster of 12 Concorde pilots in August 1999. He had 13,477 flying hours, of which 5,495 were as aircraft commander and 317 on Concorde. Marty was 54 — two more years and he could retire.

The co-pilot, Jean Marcot, 50, was also a record-holder — in 1992 he was a member of the Concorde crew that flew around the world in 32hr 49min, the fastest time ever. With a Concorde rating since 1989, Marcot had often been offered promotion to the captain's seat on other types of aircraft. But he loved flying Concorde and told friends that he would never go back to 'driving vans'.

On Tuesday 25 July Marty and Marcot began flight planning as their passengers arrived in Departure Hall 2A in Terminal 2. Air France had placed pot plants to cordon off check-in No 6, which was being used by Concorde passengers. There were delays and some of the passengers were late. Luggage had to be transferred from feeder airlines and passengers had to pass through French customs.

On scanning Tango Sierra Charlie's engineering log, Marty saw that it had returned from New York the previous day with a faulty thrust-reverser mechanism on No 2 (port inner) engine. No replacement part had been available, but maintenance supervisors had declared the problem to be *'en tolérance technique'*, meaning Tango Sierra Charlie could still be flown within the manufacturer's safety parameters.

Maximum Concorde take-off weight is 185.070kg. Marty knew that every seat on his flight was booked and, judging by the cargo weight, the German tourists were bringing plenty of luggage. With the projected take-off weight at the limit, Marty knew that if he had to do an emergency stop at Charles de Gaulle, he would need all four thrust reversers to halt the heavily laden airliner.

Left:
Air France
Concorde in
all its glory.
The Aviation
Picture Library

11

The problem lay not with the thrust reverser itself but with the hydraulic pump that operated it — a part that was easily accessible on top of the wing above the engine. Given that he had over an hour before take-off — the scheduled local time of departure was 16.25hrs — it is not surprising that Marty decided that there was time to do the repair, especially given the passenger delays. A van was dispatched to a hangar where an old Concorde sat surrounded by scaffolding; a hangar queen waiting to be pillaged for parts. A two-man maintenance crew returned with the required pump and set to work. In 30 minutes it was fitted and the repairs were completed to the flight crew's satisfaction.

Paris Charles de Gaulle has one northern runway, 09/27, and two southern parallel runways, 08/26. Work was being carried out on 09/27, so AF4590 was directed to leave from Runway 26 Right. At 16.07hrs, ATC gave start-up clearance and confirmed 26 Right for take-off. Squeezed into Concorde's cramped cockpit, Marty and Marcot ran through their final navigation and radio equipment checks. Behind them the flight engineer, Gilles Jardinot, aged 58 with 12,532 flying hours, of which 937 were on Concorde, checked hydraulics, air conditioning and electronics. It was a reflection of Tango Sierra Charlie's high all-up weight that, on being told to plan for a take-off from Runway 26 Right, the crew told ATC that 'we need the whole length of 26 Right'. At 16.22:22, Capt Marty confirmed that 'we're going to be at the . . . structural limits'.

At 16.34hrs Ground Control cleared the aircraft to taxi towards the 26 Right (26R) holding point via the Romeo taxiway. Two of the engines were started while Concorde was pushed back. As Air France Four Five Nine Zero taxied away from the terminal, a further 30 checks were made of brakes, controls, centre of gravity and so on. There were no problems, so Jardinot fired up the other two engines. From down the back, Cabin Services Director Huguette Le Gouadec from Brittany reported that she and her five flight attendants had completed safety announcements and that all passenger seatbelts were fastened.

At 16.39:04secs Capt Marty gave his take-off brief. 'So the take-off is . . . at maximum take-off weight, one hundred eighty tonnes . . . between one hundred knots and V1, I ignore the gong. I stop for an engine fire, a tyre flash and the failure callout. After V1, we continue on the SID [Standard Instrument Departure] and we land back on Runway Twenty-six Right.' At 16.40hrs, Tower cleared Four Five Nine Zero to line up. From his seat on the flightdeck, Capt Marty could see all 3,370m of 26R stretching into the cornfields and blue sky. The sun was shining, the weather fair and New York was just over three hours away.

At 16.42:02 Four Five Nine Zero was given take-off clearance with a wind of 8kt at 090°. The crew read back the take-off clearance. Flight Engineer Jardinot stated that the aircraft had used 800kg of fuel during taxying. At 16.42:31 Capt Marty began his take-off roll, holding the throttles open with his right hand while keeping a firm grip on the control column with his left. The President of France, M Jacques Chirac, who had just landed in a Boeing 747 after returning from a G8 summit in Tokyo, watched from the window of his aircraft as Concorde began its run. Twenty-three seconds later, the first officer called 100 knots, then V1 (150kt) nine seconds later. From now on, there was nowhere to go but onward. Given the outside air temperature of 19°C and a surface wind fluctuating between calm and

9kt, and between 330° and 170° in direction, there was not enough runway left for Concorde to stop on.

At 16.43:13, as Marty rotated Tango Sierra Charlie at 198kt for lift-off, the Tower controller cried out '. . . Four Five Nine Zero you have flames . . . you have flames behind you.' The first officer acknowledged this as the flight engineer announced the failure of No 2 engine. The flight data recorder subsequently showed a momentary loss of power on No 1 engine, but this was not mentioned by the crew, who probably had enough to be going on with. Eight seconds later the fire alarm sounded and Jardinot announced that he was shutting down No 2 engine. The fire alarm then stopped. The first officer told Capt Marty, 'Watch the airspeed, the airspeed, the airspeed.'

Marty, who once flew a hang glider over a smoking volcano in Guadeloupe, appeared to remain calm. Although the port inner engine was now shut down, the silver bird could still fly on the other three. In a well-rehearsed drill the flight engineer would have flicked a bar of white switches at the base of the throttles to get 'contingency power'. Fuel was already ramming into the Olympus engines at eight gallons a second, generating temperatures above 1,000°C. The addition of an extra 7% thrust threw more fuel on to the flames.

At 16.43:28 someone on the flightdeck was heard to say, 'It's really burning and I'm not sure it's coming from the engines.' Two seconds later the flying pilot called for landing gear retraction. The controller confirmed, 'Four Five Nine Zero you have strong flames behind you [and] as you wish, you have priority for a return to the field.' The crew acknowledged this transmission. At 16.43:42 the fire alarm started up again.

At 16.43:56 the first officer reported that the landing gear had not retracted and made several references to the airspeed. At 16.43:59 the Ground Proximity Warning Alarm sounded several times. At 16.44:05 the controller transmitted, 'Fire Service Leader er . . . the Concorde, I don't know his intentions, get into position near the southern parallel runway.' This was followed by, 'Fire Service Leader, correction, the Concorde is returning on Runway 09 in the opposite direction.' The flight crew then transmitted, 'We're trying for Le Bourget . . .'. The flight data recorder then indicated loss of power on engine No 1.

Le Bourget is a smaller airfield a few miles away from Charles de Gaulle. As he grappled with the controls, Marty tried to gain height but his Concorde never got higher than 200ft above the ground. It seemed to lack the power to climb. A few seconds later, Tango Sierra Charlie crashed on to a hotel at La Patte d'Oie in Gonesse at the intersection of the N17 and D902 roads. The cockpit voice recorder (CVR) stopped recording at 16.44:31. At 16.46:09 the controller announced, 'For all aircraft listening I will call you back shortly. We're going to get ourselves together and we're going to recommence take-offs.' At 16.55:47 an aircraft crew told the controller, '. . . there is smoke on Runway 26 Right, there's something burning apparently, for information'. Two minutes later, a runway inspection vehicle told the controller, 'There's pieces of tyre which are burning.'

Normally, there are few if any witnesses to the unfolding of an air accident, but this was not so in the case of Air France Four Five Nine Zero. From his glass-walled office overlooking the runways at Charles de Gaulle, Jean-Cyril Spinetta,

Chief Executive of Air France, thought it prudent to watch President Chirac's aircraft land. He could not resist lingering to watch Concorde Tango Sierra Charlie move on to the runway. 'No matter how many times you've seen it, you always stop to watch. It's that kind of plane,' he told friends afterwards. Through the reinforced glass, Spinetta heard the muffled roar of the four engines powering to maximum thrust. The sleek airliner began moving forward and as the 56-year-old Corsican was about to turn away, he noticed a streak of flame burst from the left wing, followed by billowing smoke. He stared transfixed, as the streak became a fiery sheet, chasing after Concorde like an orange shroud.

Darren Atkins, a British businessman on board an aircraft waiting for clearance to take off after Concorde, had an even closer view. 'As the aircraft drew level with us — before it started to take off — the engines were burning very heavily. On the tarmac was some debris.' This was later identified as part of a tyre — at least one of the tyres bearing the 180-tonne load at 200mph had blown.

Andras Kisgergely, a 20-year-old student from Budapest, was aircraft-spotting on the airport perimeter with a friend. He raised his camera and had time to take one shot: it showed the crippled aircraft low over the trees trailing flames as long as the fuselage. As other photographs confirmed, most of the flames appeared to be coming not from the engine but the wing, which housed six fuel tanks.

Concorde Tango Sierra Charlie was now vainly trying to climb with two afterburners pushing out around 40 tonnes of thrust on the starboard side and little or nothing on the left. The main runway at Charles de Gaulle sends aircraft directly over the town of Gonesse, on the outskirts of Paris, and the instrument panel in front of Marty and Marcot would have been flashing like a pin table as Four Five Nine Zero powered towards the black and yellow building of the local hospital. Standing on the N17 road, Jacques Messier was convinced the aircraft banked to avoid the town and the hospital. This is probably wishful thinking because no one on that flightdeck would have been looking at ground features. The engines were in control and as Jean Paul Canto watched under the flight path, he saw the Concorde head for an 80m-high hill covered with lime and birch trees. It stands between Charles de Gaulle airport and Le Bourget, and Canto noted that 'he [the pilot] was heading in the direction of the hill, but he was too low. Suddenly he veered left.' Sid Hare, an American pilot, saw the aircraft rear up steeply. 'The nose pitched up, almost straight, in the vertical, that much I know for sure,' he said. 'After that I'm not sure if it just started to backslide, or continued to fall over on the left side.'

In the manager's office of the hotel Les Relais Bleus, beside the N17, Patrick Tesse was talking on the telephone when the noise outside became too loud to ignore. 'I looked out of the window and saw a burning aircraft heading straight for me. I was stupefied. I dropped the phone.' He escaped as the shrieking mass of metal and kerosene narrowly missed his hotel. The neighbouring Hotelissimo was not so lucky. Inside, a small group of staff were preparing for the arrival of the Suffolk Youth Wind Band with its 45 teenage musicians. The hotel manageress, Mme Fricheteau, 46, was also wondering about the roar. 'I said to my trainee receptionist, "Concorde is really overdoing it today, it's making more noise than usual." I had barely finished saying that when I was hit by a burst of flames in my face.'

By now the shape that had graced the skies with such elegance had become a flying bomb, with 96 tonnes of fuel encasing two banks of 50 seats. Inside the cabin, 100 passengers expecting a unique flying experience would have been gripping their seats in rising panic as the world tumbled outside the tiny portholes. The airliner impacted like a 'mini atomic bomb', according to one witness, as its tanks full of high-octane fuel ignited.

On impact the intensity of the high-temperature fireball was such that plastic parts of the neighbouring hotel melted together. Alerted by a fireman, the brigade from the south fire station at Paris Charles de Gaulle aerodrome immediately set out. At the same time, the crash alarm was activated via the local network by the controllers on duty at the southern lookout post. Firemen from Le Bourget aerodrome were first to arrive at the scene of the catastrophe, eight minutes later. Faced with the scale of the fire, they were able only to limit the fire and provide aid to the injured.

The crew were all found in their take-off positions and the passengers in the seats assigned to them at boarding. The official investigation reported tersely that 'the circumstances of the accident and the condition of the aircraft meant that the accident was not survivable'. Trapped inside with no chance of escape, the passengers would have known mercifully little. Most of the fuselage and wings were vapourised in the enormous heat of the explosion. Only the pilots may

Above:
Concorde in flames
at lift-off.
Associated Press

have known for a split second what was coming as the Concorde seemed to die on its side, with perhaps the rear end striking the ground first. Somehow, the nose cone and part of the cockpit survived in recognisable pieces despite the ferocity of the explosion.

Given that the hotel was almost entirely flattened, it is miraculous that only four people died in it and only another six were injured on the ground, two chambermaids — one from Mauritius with two young children — and two Polish helpers being engulfed by flames inside the Hotelissimo. Many of the bodies were unrecognisable. As rescuers, including 40 forensic experts from a Catastrophe Victim Identification Unit, began the grisly task of sifting through the wreckage, they marked the site of human remains with traffic cones.

Of the 100 passengers who perished, three generations of the Eich family had been wiped out. Six families from Mönchengladbach had been devastated, losing 13 people, including an eight-year-old child. Eight victims came from Berlin; another 26 from North Rhine-Westphalia; 11 from Hessen; 10 from Bavaria; seven from Hamburg. German Chancellor Gerhard Schröder led the prayers in a memorial service in Hannover for the dead. President Chirac led a similar service at Gonesse.

*   *   *   *   *

Being unsure of the cause of the crash, Air France and British Airways suspended all Concorde flights. The investigation into why the supersonic delta-winged airliner came to such spectacular grief was given under French law to teams from the Gendarmerie des Transports Aériens and the Bureau Enquêtes-Accidents (BEA), working with British air accident investigators and aerospace engineers from industry.

Examination of the site showed that the aircraft had struck the ground about 9.5km from the threshold of Runway 26R with little horizontal speed. After the impact, it broke and spread generally to the south, with the aircraft upright.

Only the front parts of the aircraft escaped the ground fire, together with a few pieces of the fuselage scattered over the site. Most of the wreckage, with the exception of the cockpit, remained within a rectangle measuring 100m long by 50m wide. The outer part of the left wing, with the outer elevons still attached, was found melted on the ground. Nearby was the inner part of the wing with Nos 1 and 2 engines still attached. The engines were resting on a water tank 1.5m in height. Many wing parts were found nearby, and the lower parts of the left and right main landing gears were close to the initial impact marks. The landing gear was down and locked at the time of impact.

A large number of parts belonging to the cockpit hit an electric power transformer. The pilots' seats, the throttle levers and the autopilot control unit were here, and the seven landing gear ground lock pins were found with their stowage bag. Next to this there was a section of the fuselage aisle between cockpit and cabin. Nearby, the nose landing gear was found, extended.

The main components of the Concorde's structure were found along the axis of the wreckage scatter. The passenger cabin was identifiable from pieces of fuselage, together with a large number of items of debris from the hotel.

Some 100,000 Concorde fragments were brought to Hangar 12 at Dugny military base near Le Bourget airport. The jagged chunk of white fuselage bearing the letters 'NCE' of the blue Air France logo and the blackened seat cushions losing their stuffing were a world removed from any previous glamour. Piles of what looked like junk, the biggest piece of which measured barely 3-4m, were sifted and those considered vital to the investigation were sent away to laboratories for analysis before being returned to a second hangar where Tango Sierra Charlie was partially reconstructed. The two port engines that failed during take-off were examined by experts at the Centre for Propulsion Studies in the Parisian suburb of Sanglay. Thousands of tiny bits of metal and debris, destroyed beyond recognition and impossible to identify, were placed in red boxes or plastic bags in the centre of the hangar. Nothing could be discarded until the inquiry was completed.

It soon became clear that the disaster was not caused by a bomb. The engineers responsible for the last-minute repair to No 2 engine thrust reverser were interviewed, but by the end of the first week attention focused on the fact that one, and possibly two, of the Concorde's Goodyear tyres had burst.

Like good detectives, the accident investigators were greatly helped by following the trail of debris and marks left on the runway and under the aircraft's flight path. Working out from the 26 Right threshold, the first evidence was parts of the water deflector from the left main landing gear. A deflector is located at the front of each main landing gear to deflect water and spray away from the engine air intakes, and they were made from glass fibre and composite materials to make them frangible.

Next came the pieces of Concorde tyre. One piece measured 100cm x 33cm and weighed more than 4kg. When the pieces were fitted together, visual inspection revealed a transverse cut about 32cm long.

A strip of metal about 43cm long, bent at one of its ends, was then found on the runway shoulder. Its width varied from 29 to 34mm and it had drilled holes, some containing rivets. The piece appeared to be made of light alloy, coated on one side with greenish epoxy primer and on the other side with what seemed to be red aircraft mastic for hot sections. It did not appear to have been exposed to high temperature. This 43cm-long strip of metal was not part of the Concorde, and it was alleged that it had fallen from a Continental Airlines DC-10 that took off on 26R two minutes before Air France Four Five Nine Zero.

Then came signs of an explosion and a piece of concrete torn from the runway, followed by small pieces of broken runway light some 2,800m from the runway threshold. Ground marks showed these to have been caused by the Concorde's left main landing gear. After this point, the marks became intermittent as the Concorde staggered to become airborne. From 1,807m to 2,340m the mark of a flat tyre with an incomplete tread was observed. When the mark disappeared at about 2,340m, its displacement from the centreline was about 8m. This corresponded to the right front tyre of the aircraft's left landing gear.

Amateur video of the tragedy showed Tango Sierra Charlie trailing flames from its left side while the aircraft was accelerating between V1 and VR. As this area of the left wing was not provided with extinguishing equipment, an inferno soon enveloped the port underside. The grass was burnt adjacent to the runway

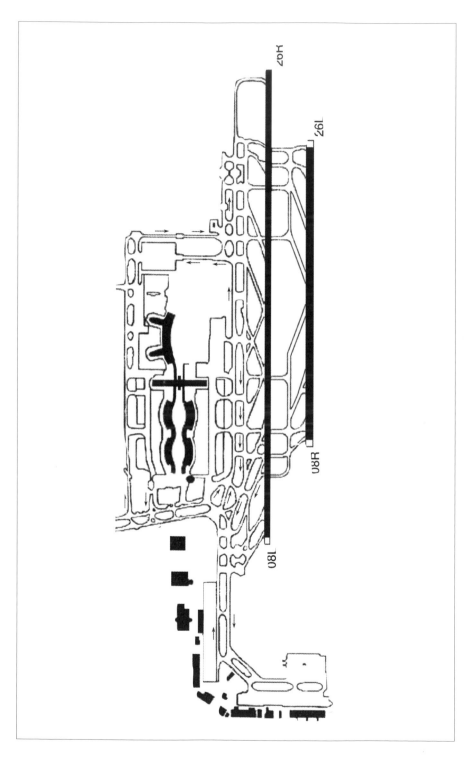

edge between 2,902m and 3,165m — this area, with its soot deposits, showed that there had been extensive flame after the Concorde became airborne. Thereafter an almost continuous trail of debris stretched to the impact point. Burn marks were visible on the ground where certain items of debris were found, and a wheat field was damaged by fire 2,500m beyond the end of the runway.

From all the evidence, witness statements and photographs, BEA accident investigators concluded that during Four Five Nine Zero's take-off run, the front right tyre of the left main landing gear ran over the errant strip of metal. Between V1, which was reached 1,150m or 33 seconds after brake release, and Rotate Speed (2,070m or 48 seconds), the shredded tyre shot chunks of rubber into the underside of the aircraft's wing, rupturing at least one fuel tank. Initial ignition of the major fuel leak began under the wing, between the left engine nacelles and the fuselage. A few seconds before take-off rotation, a small flame appeared suddenly, similar to a blowtorch, and then widened to envelop the left engines and run down the fuselage. This flame was accompanied by thick black smoke, and it turned into such a huge fire, accompanied by loss of thrust on one, and then two engines, that Capt Marty was unable to stabilise the stricken airliner. With three operating engines and the gear stuck down, Tango Sierra Charlie stopped climbing. Marty needed a flying speed of 205kt just to keep level, and over 300kt with two engines. The highest speed achieved by Four Five Nine Zero was 211kt before power on No 1 engine began to run right down. Thereafter the speed dropped until the Concorde became unflyable.

Tango Sierra Charlie flew over the N17 road at about 200ft, turned to the left at a steep bank angle, pitched nose up and crashed left wing first. The Concorde crashed less than 90 seconds after the destruction of the tyre. Investigators concluded that the Air France crew would have had no way of knowing the extent of the fire, nor could they have taken any action to contain it.

Each of Concorde's two main landing gears consists of two four-wheel bogies. The bogies are equipped with a system that detects under-inflation of a tyre and transmits a visual signal to the cockpit, but this detection system is inhibited when the indicated airspeed exceeds 135kt. Unfortunately for Capt Marty and his crew, it seems that Tango Sierra Charlie was accelerating away from 135kt when the right front tyre on the left main landing gear was shredded.

Concorde tyres take such a hammering that British Airways changes them approximately every 25 flights, compared with about 100 or more flights on an ordinary jet. The shredded tyre on Tango Sierra Charlie was on its 42nd flight, which was one reason why French accident investigators identified what they called 'various problems' with Air France's operation of the aircraft. One 'serious error' by maintenance staff was the failure during maintenance shortly before the crash to replace a bracing strut in the left-hand undercarriage bogie. This meant that there was no spacer between the Concorde's wheels, which would have caused it to wobble and veer to the left during its take-off run, slowing down the crippled aircraft and worsening its plight. French investigators stated that 'the fact that this negligence did not contribute to the accident does not make it any less serious'.

Concorde tyre bursts or deflations happen at the rate

Left:
Runway layout at
Paris Charles de
Gaulle airport.

19

of one for every 4,000 flying hours — about 60 times more frequently than the rate for subsonic long-haul aircraft such as the Airbus A340 — which explains why an under-inflation warning system was incorporated to alert the pilots. Moreover, not only was tyre damage during taxi, take-off or landing not unknown on Concorde but also it had previously led to damaged structure and systems. Crash investigators found that of 57 cases of Concorde tyre bursts, 30 occurred on aircraft operated by Air France and 27 on the British Airways fleet. In 12 instances wings or tanks incurred structural damage, and in six cases fuel tanks were penetrated.

Debris from the tyre and shrapnel from the wheel had punctured three fuel tanks on 25 July, causing shock waves to rupture the tank from within. Two consequences were a damaged No 2 engine and a large hole torn in the top wing skin, but Capt Marty could have coped with those. Tango Sierra Charlie was brought down by the fact that escaping fuel was whisked by the turbulence around the landing gear and caught fire. This blaze could not be extinguished while the airliner could neither climb nor accelerate. The problem for future Concorde operations was that no one knew for certain the cause of this combustion.

Since Concorde entered service, six cases of damage to tanks had been recorded but none of them had resulted in a fuel fire. As the disaster on 25 July 2000 showed that the destruction of a tyre — a simple event which could not be prevented from recurring — had catastrophic consequences in a very short time-scale without the crew being able to recover from this situation, French and British authorities agreed to suspend the Concorde Certificates of Airworthiness 'until appropriate measures have been taken to ensure a satisfactory level of safety as far as the tyre destruction based risk is concerned'. In other words, until investigations had identified the reason why fuel escaping from Air France Four Five Nine Zero caught fire as quickly and as dramatically as it did on 25 July, Concorde would not be allowed to resume commercial operations until modifications were put in place to stop any fire from having the same awful consequences in future.

The most important of the resulting seven modifications incorporated on Air France's five Concordes and British Airways' seven were to add Kevlar liners to all or part of the six most crucial fuel tanks, to change to newly developed Michelin tyres on the main landing gear and to strengthen the undercarriages. Bulletproof Kevlar linings are commonly used to protect fuel tanks on military helicopters and Formula One racing cars, and they reduce the rate of a fuel leak from 100 litres a second, as in the Paris crash, to 1 litre a second. Although a Kevlar lining is the thickness of a credit card, the weight of lining six fuel tanks meant that the cabin interior had to be fitted with lighter seats. Concorde's new Michelin tyres were much less likely to burst and, if they did, would break into much smaller pieces. British Airways alone spent £17 million on the technical improvements and a further £14 million on a revamped cabin and other enhancements.

French and British Concordes returned to commercial service on 7 November 2001. Showbusiness personalities posed for the tabloids as less famous guests sipped champagne and ate sea bass 16km high. The two airliners landed in New York within minutes of each other and 'touched noses' on the tarmac at JFK in a symbolic gesture of international unity. New York was overjoyed at the reunion, less than two months after the terrorist attacks on the twin towers of the World

Trade Center. The city's mayor, Rudolph Giuliani, used the occasion to jump aboard BA's Alpha Echo to welcome it back. Within hours of the first commercial flight departing from Heathrow, Prime Minister Tony Blair was doing his bit for Britain by taking a second Concorde for a two-hour meeting with President Bush in Washington.

When the British Concorde landed in New York on 7 November after 15 months of enforced idleness, Capt Mike Bannister told passengers over the cabin's intercom: 'We have put Concorde back where she belongs. Sit back and relax. We're glad to be back.'

*   *   *   *   *

At any time there are around 1,200 Boeing airliners in the world's skies, compared to about three Concordes flying to New York and back. The dozen or so Concordes put together have flown a total of about 80,000 flights in the last 25 years — in contrast, the world's most widely used aircraft, the short-haul Boeing 737, makes 80,000-plus flights every week.

This is worth bearing in mind when proclaiming that Concorde enjoyed 25 years of accident-free flying up to 25 July 2000. French accident investigators indicated that the apparently fatal 43cm strip of metal found on the runway appeared to have been attached to a DC-10 during repairs in Houston, 16 days before the Concorde disaster. 'The level of wear on the strip adjacent to the missing strip was greatly in excess of the tolerance permitted by the manufacturer,' a BEA report stated. 'Various questions' about the maintenance of the DC-10 remain to be answered, but bits of metal and stones are swept up from airport operating surfaces every day. Perhaps the sleek delta-winged Concorde was too delicate for the regular rough and tumble of scheduled operations before 25 July, and it is now safer because it has been beefed up?

Supersonic airliners do not deserve any special treatment. After the Paris crash, British Airways introduced extra runway inspections immediately before each take-off during the three weeks before its seven Concordes had their airworthiness certificates withdrawn. The checks proved to be both time-consuming and difficult to co-ordinate because of the intensive use of Heathrow's two runways. Heathrow air traffic personnel inspect the runways four times a day. More than 250 aircraft take off and land between each inspection, and conventional airline operators cope with this balance between an oppressive safety regime and commercial throughput.

The sensible answer to any major flight safety problem is not to shut the stable door after the horse has bolted, but to prevent the steed from ever escaping again. Consequently, Air France and British Airways abandoned any suggestion of checking the runway before each Concorde departure in preference to fitting new tyres designed to resist any sharp objects that might lie around.

The French accident report exonerated Capt Marty's crew of any error in their procedures after the fire broke out on the runway. It concluded that the destruction of the aircraft was inevitable, given the fury of the fire, whether the captain had decided to abort the take-off or proceed with it, as he did. However, investigators identified what they termed 'improvisation and lack of method' which could have affected the aircraft's operation. Apart from the bracing strut missing from the left undercarriage bogie, passenger baggage was mishandled and Tango Sierra Charlie was allowed to take off when overloaded by a tonne and with a stronger-than-expected tail wind. Although this had no practical impact on the aircraft's performance, investigators identified a 'strong-willed can-do' culture among the elite Concorde pilots rather than the meticulous attitude required of *all* flight crew. It was symptomatic of the culture that Concorde pilots were a breed apart from lesser aircrew that the Concorde's first officer's licence was found to have legally expired the previous week because he had not got round to updating the medical certificate required to validate it.

Concorde returned to Air France and British Airways service not so much for the revenue as for the elegant, futuristic cachet that was seen as a key part of an upmarket image even after a quarter century in service. But perhaps it took that long to prove that, notwithstanding the cutting-edge technology, the most advanced aeronautical icon can still be brought down by old-fashioned causes.

2
Fire Up Above

A Sikorsky S-76 helicopter was just about to taxi when the aft baggage bay smoke caption illuminated. The bay was inspected but nothing untoward was found. During the subsequent flight the smoke caption continued to illuminate intermittently. After landing, a mobile phone inside a bag was found to be switched on. It was switched off and subsequent legs were flown with no recurrence of the problem. Tests confirmed that a live mobile phone could illuminate the smoke caption.

But generally there is no humorous side to a fire alarm in the air. Eric Hilliard Nelson was born on 8 May 1940. By the age of eight, Eric had become 'Ricky' and had joined his family in a popular radio series called *The Adventures of Ozzie and Harriet*. The show moved to television and Ricky's good looks and media-wise personality gave him a head start when he decided to cross over to pop music in 1957. A string of hits followed his million-selling release 'I'm Walking' including 'Poor Little Fool', 'Travelin' Man' and Gene Pitney's 'Hello Mary Lou'.

Notwithstanding roles in the movies, Ricky's chart appearances began to tail off in the Beatles era. He left the pop world for country music, becoming 'Rick' in the process, and by 1985 he had assembled a new, vibrant band that toured extensively. For all of his life, Rick Nelson had had a great fear of flying, such that he insisted on always flying commercially, and never in anything with a propeller. He broke both of these rules when he decided to buy N711Y, a vintage Douglas DC-3 previously owned by the piano-pumping rock-and-roller Jerry Lee Lewis. N711 was dubbed 'the flying bus' because of its sluggishness and tendency to malfunction on the runway. The temperamental gas heater in the passenger cabin had often to be adjusted in mid-flight by the pilot when the cabin got too cold for the weary band.

On Tuesday 31 December 1985, Nelson's DC-3 was *en route* from Alabama to a New Year's Eve show in Dallas. At 17.08:48hrs, while cruising at 6,000ft, one of the two pilots of N711Y advised ATC, 'I think I'd like to turn around, head for Texarkana here, I've got a little problem.' He was given a vector and advised of the closest airports. Shortly afterwards, he stated he would be unable to reach any of the airports, and at 17.11:49hrs he reported smoke in the cockpit. After crash-landing in a field near De Kalb, Texas, at 17.14hrs, the DC-3 hit wires and a pole before continuing into trees, where it was extensively damaged by both impact and fire. The crew got out through the cockpit windows but the seven passengers were not so fortunate.

Rick Nelson died at the age of 45, a household name and former teen idol with 18 top 10 singles and who was ranked among the top half-dozen singles sellers of all time. But for all his money and fame, he ended up on a transport aircraft built in 1944 which was so antiquated that, notwithstanding repeated attempts by the captain, it proved impossible to start the cabin heater in mid-winter. At some stage on the flight to Dallas, smoke entered the cabin and, despite the opening of the fresh air vents and the cockpit windows, became so dense that the pilots had difficulty seeing.

After the crash, it was found that neither the oxygen system nor the hand-held fire extinguishers had been used. Investigators determined conclusively that the fire originated on the right-hand side of the aft cabin area, at or near the floor line. The fire had begun in a malfunctioning gas heater, and the heater door was found to be unfastened. The resulting pungent smoke was so choking that none of the passengers was able to get out of the wreck.

A superior pilot is someone who never allows himself to get into a situation that demands the use of his superior skills. One such aviator was Sqn Ldr Hedley George Hazelden, who survived the rigours and risks of nightly bomber operations throughout World War 2 before being sent on the first course at Britain's Empire Test Pilots' School. After the war 'Hazel' became chief test pilot at Handley Page, where most notably he carried out flight test development of the Hastings military transport, the Victor four-jet nuclear bomber and tanker, and the Hermes and Herald airliners.

The twin turboprop Herald was a small airliner designed to take the place of the DC-3 Dakota. On 10 August 1958, Hazel was piloting one to demonstrate its potential at the Farnborough Air Show. On the way there, the Herald's starboard engine caught fire. Hazel cut off the engine's fuel supply, feathered the propeller and pressed the fire extinguisher button, but the blaze continued. With eight passengers on board, including his wife, he had no alternative but to make an immediate forced landing.

All he could see ahead was heavily wooded countryside, so he throttled back the remaining engine and descended until an open field came into view. The starboard tailplane had already burned away, and now the Herald's starboard engine fell off, making the aircraft all but uncontrollable. Approaching the field, Hazelden spotted an 80ft tree in front of the only possible place to touch down. Worse, a farm roller was parked across the touchdown area and high-tension cables crossed the ground he needed for his after-landing roll.

In an astonishing manœuvre, with his right wing now well on fire, Hazelden flew over the tree, belly-landed and slid under the wires. A concealed tree stump tore a hole in the fuselage in front of the burning wing, which fortuitously provided an escape route for all on board before the Herald burned out.

Hazel survived probably because he trained in an age that emphasised the importance of the rudder. Close to the ground, be it in formation in the wake of another aircraft, when facing wind shear or, as in Hazel's case, when he was losing both right elevator and right aileron, power and rudder were all he had left to control direction and keep the wings level. Sqn Ldr Hazelden was such a good pilot that he died in his bed in August 2001 at the age of 86. In the course of decades of operational and test flying, on more than one occasion he saved his own and his crews' necks by dint of his superb airmanship. They don't make airmen like that any more, not least because accountants rarely see the value-added in putting their pilots through the sort of flying exercises old hands used to learn in air forces around the world.

Fire in the air has always been an aviator's nightmare but to put the risk in perspective, cabin fires are extremely rare in passenger airliners. Unfortunately, when they do occur their effects can be devastating, which explains why national aviation authorities act promptly when post-fire lessons have to be learned. For example, two airliner toilet fires in the summer of 1974, both fortunately non-fatal, prompted the installation of automatic-discharge fire extinguishers in washroom waste paper containers in all airliners and the prohibition of smoking in lavatories. On 2 June 1983, an Air Canada DC-9 was cruising from Dallas to Toronto at FL330 when a fire started in a lavatory, probably caused by a motor overheating. The passenger cabin filled with smoke so the crew declared an emergency, but it took about 20 minutes before they could land the DC-9 at the closest large airport, near Cincinnati. When the airliner finally stopped, only half of the 46 people on board were able to escape from the burning aircraft before being overcome by smoke and fumes. As a result, a number of safety recommendations were introduced both to reduce the likelihood of an in-flight fire and to slow its progress if it did develop. Out of this tragedy came the fitment of smoke detectors and automatic-discharge fire extinguishers in lavatories, stiff fines for those who attempt to disable a smoke detector, floor-level escape lighting along aisles to guide passengers toward an emergency exit should visibility be reduced by smoke, and the requirement to fit all airliners with fire-blocking cabin and seat materials.

Cargo compartment fires became the next fire safety concern. In 1981, a Saudi Arabian Airlines Lockheed L-1011 experienced an in-flight fire after departing Riyadh for Jeddah. Although the widebody returned and landed at Riyadh, all 301 persons aboard perished in the ensuing cabin fire, which was later found to have started in the aft cargo compartment. Subsequent safety recommendations led to major modifications of all widebody aircraft cargo compartments to prevent a fire from spreading. But, as the eminent German playwright Schiller wrote two centuries ago, 'Against stupidity the gods themselves struggle in vain.' On 2 February 1988, an American Airlines DC-9 landed safely at Nashville with a burning cargo compartment, caused by an improperly packaged and prohibited chemical shipment. After this incident, the fitment of fire detection and suppression equipment was recommended in cargo areas that were previously thought to be so airtight that a fire could not be sustained. But no regulations were issued, and then on 11 May 1996 a ValuJet DC-9 crashed in Miami after a fire erupted in a cargo compartment. The NTSB investigation concluded that activation of one or more chemical oxygen generators in the forward cargo compartment initiated the fire. In early 1998 a rule was issued requiring fitment of fire detection and suppression systems in the cargo holds of 3,700 aircraft by 2001.

Through a mixture of regulatory pressure, enhanced fire warning systems, fireproofing and passenger education, the risk and impact of cabin fires have been greatly reduced over the past 20 years. The main risk from fire in the air today, albeit relatively small given the millions of hours flown annually, comes from failing equipment operating at high temperatures. And, like Rick Nelson's elderly DC-3 cabin heater, long-serving engines and motors need careful watching if they soldier on long after their designers ever expected them to.

In days of yore, when flying was dangerous and sex was safe, professional soccer players earned a basic wage and had an unpretentious lifestyle to match. Nowadays, life in the Premier League is so frenetic and glamorous that highly paid stars often travel to and from an away game by charter air. So it was on 30 March 1998, when the Leeds United team flew home after a match at West Ham in east London. They were booked on an Emerald Airways Hawker Siddeley 748 turbo-prop (G-OJEM) scheduled to depart from London Stansted Airport at 22.30hrs on a one-hour flight to Leeds Bradford Airport. A baggage problem delayed the flight and the aircraft eventually taxied at 23.23hrs to the holding point for Runway 23. Take-off clearance was given at 23.29hrs. The HS748 had a water/methanol injection system, which, in the event of loss of thrust from an engine on take-off, automatically selected full 'wet power' on the other engine. First Officer Garry Lucas, aged 33, was the handling pilot and, as the take-off was to be made with full dry power, the water methanol system was selected to standby.

With the throttles wide open, First Officer Lucas confirmed that the warning lights were out and the emergency panel was clear. As the HS748 accelerated, the aircraft commander, Capt John Hackett, announced 'sixty knots' and relinquished steering control to the first officer, who acknowledged and confirmed, 'full dry we have, just slightly low on the right'. No significant variation in rpm between the two engines was noted by the flight data recorder. At 111kt, when the Commander called 'V1, rotate', the first officer pulled the control column rearwards and G-OJEM got airborne.

Less than five seconds after the 'rotate' call, at 115kt and 30ft-100ft above the ground, there was a sharp report followed by an engine run-down. The turboprop yawed 11° to the right of the runway heading and, as the pilots asked each other what the noise had been, loud shouting could be heard from the passenger cabin. The first officer said, as he corrected the yaw, 'something's gone' and Capt Hackett took control of the aircraft. While the first officer noted that an engine had stopped, Senior Cabin Attendant Helen Dutton, 33, told the 40 passengers to sit down before advising the flightdeck crew via the interphone that the right engine was on fire. Engine power was reduced and the aircraft yawed 14.5° to the left of runway heading. Four seconds later, the sound of the engine fire warning bell sounded. Helen Dutton told the passengers to 'stay in your seats and make sure your seatbelts are all fastened'.

Faced with what appeared to be an uncontained failure of the right engine, resulting in sudden power loss and a major engine bay fire, Capt Hackett elected to throw the HS748 back on the runway. G-OJEM had been airborne for all of 27 seconds before it was back on the ground. The commander called for brakes, to which the first officer replied 'coming on', but given the limited runway left when Capt Hackett put the HS748 down, no one was surprised when G-OJEM left the asphalt doing around 62kt.

G-OJEM came to rest in the overrun area beyond the end of Runway 23, within the airport boundary. Tyre marks showed that after leaving the runway, the HS748

continued to roll straight on until it encountered the road, 78m after the end of the runway. At this point the tyre marks associated with all three landing gears ceased. Main gear tyre contact was re-established on the second half of the road, but the signs were that the nosewheel tyres did not re-contact the ground until 18m beyond the road leading edge. The marks and wreckage distribution showed that the nose landing gear, which was not built for off-road travel, collapsed rearwards at this point. The HS748 slid for a further 22m on its forward fuselage undersurface and mainwheels before coming to rest 118m beyond the end of the runway, 558m from the initial touchdown point. The flight data and cockpit voice recorders stopped with the loss of electrical power 7.1 seconds after the aircraft left the paved surface.

After G-OJEM came to a halt, First Officer Lucas immediately left his seat to assist with the evacuation of the aircraft. The commander shut the HP (High Pressure) cocks and fuel pumps, checked the water methanol system was off and closed down the electrics. He then satisfied himself that the cabin was clear before leaving the aircraft. All 40 passengers and four crew evacuated the HS748 with nothing more than a few minor injuries.

The HS748 came from the Avro design office that created the Lancaster bomber. The twin-engined airliner was approved for commercial use in 1962, with G-OJEM rolling off the production line in 1983, a year before manufacture of the type ceased. G-OJEM had 18,352 hours on the clock on 30 March 1998. At that time there were approximately 260 similar types in service carrying passengers and freight, including 14 in the UK.

Above:
Hawker Siddeley 748 G-OJEM after it came to grief at Stansted with the Leeds United team on board on 30 March 1998.
A. S. Wright

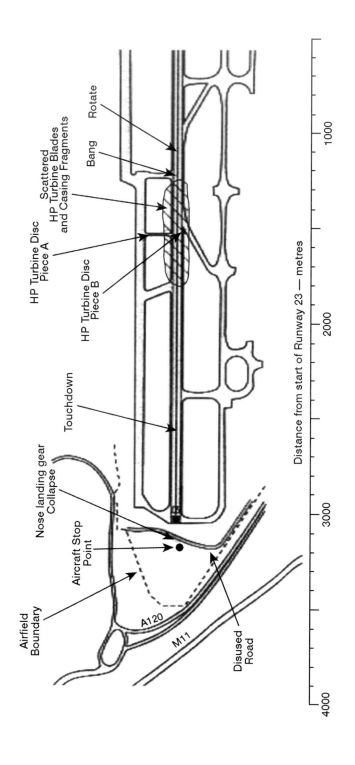

HP Turbine Disc Piece A

Scattered HP Turbine Blades and Casing Fragments

Bang   Rotate

HP Turbine Disc Piece B

Touchdown

Nose landing gear Collapse

Aircraft Stop Point

Airfield Boundary

A120

M11

Disused Road

Distance from start of Runway 23 — metres

1000

2000

3000

4000

G-OJEM was powered by two Rolls-Royce Dart gas turbine engines, located in wing-mounted nacelles and each driving a four-bladed propeller. Dart design started in 1945 and production ceased in 1987 after approximately 7,100 engines had been delivered. A significant quantity of Dart debris was found scattered between 1,250m and 1,800m from the threshold of Runway 23. This included fragments of the nozzle guide vanes, blades and disc associated with the No 2 engine HP turbine. The HP turbine disc itself was found in two major pieces, each 35-45m from the runway centreline.

The passengers seated on the right side of the cabin behind the wing reported seeing light-coloured sparks from No 2 engine jet pipe during the take-off run. Around the lift-off point, this developed into a shower of sparks and an orange glow in the jet pipe, followed by a dull bang and then flame from the top of the nacelle. The fire grew until it was described as a sheet of flame over the inboard side of the nacelle. The fire continued after the aircraft had come to rest.

Evidence from witnesses, the flight data recorder and distribution of the wreckage showed that the failure had occurred around 4 seconds after lift-off, when the aircraft was some 60ft above the ground. The failure caused an abrupt loss of power from the engine and immediately initiated a substantial fire around the engine nacelle. It also caused the automatic selection of full wet power on the No 1 engine, which gave best climb performance but also resulted in maximum power asymmetry, causing the aircraft initially to yaw 11° right.

Four similar Dart turbine failures had occurred over the previous 26 years, all attributed to a combination of turbine entry flow distortion and turbine blade wear. G-OJEM's engine failure was initially ascribed to the same causes, but after protracted study and testing by the Dart manufacturer, the evidence eventually indicated that disc fatigue strength had been significantly reduced by surface corrosion, resulting in fatigue cracking. Mandatory modification requirements to address the causes were issued in April 2001. This was not before time because another turbine failure, similar to G-OJEM's, occurred in June 2001.

Violent disruption of the engine associated with the turbine failure started a substantial overboard fuel leak from the fuel heater assembly forming part of the Low Pressure fuel supply line in the nacelle fire zone. The evidence indicated that this leaking fuel had ignited almost immediately. Although all six fire appliances in the Stansted airport fire station were at the crash site within two minutes of the initial alert call, where they put the fire out, the fuel leak itself continued for some hours, until the system was manually isolated.

It was fortunate that the HS748's passenger cabin did not fill with toxic smoke and debris, because the emergency evacuation of those inside left something to be desired. The experiences of the cabin crew and passengers during the hectic 27 seconds that G-OJEM was airborne illustrated the confusion and misunderstandings that can accompany an unusual emergency situation just after take-off. When the nacelle fire became evident shortly after rotation, several passengers started to shout. Senior Cabin Attendant Helen Dutton reported that some unfastened their seatbelts and stood up. In order to regain control of the cabin, she had to unfasten

Left:
Flight track path covering the 27 seconds G-OJEM was airborne before it ran off the end of Stansted Runway 23.

her seatbelt momentarily to get the PA handset from its stowage to tell the passengers to sit down. She then told the commander via the interphone that there was a fire in the right engine, again having to unfasten her seatbelt in order to reach the interphone controls.

There was no time to give a full emergency brief or to prepare the escape chutes, but the passengers were told to stay in their seats and ensure that their seatbelts were 'all fastened'. Unfortunately, in the confusion some passengers heard this as 'unfastened' and so they unfastened their seatbelts to adopt the brace position. Fortunately, the cabin area suffered no damage and none of the occupants was injured in the impact. Once the HS748 came to a halt after the nose landing gear collapsed, everyone was ordered out on the left side to keep away from the fire burning around the starboard nacelle. The passengers seated by the left over-wing exit had no problems opening it, and evacuation started almost immediately under the guidance of the second cabin attendant. While Capt Hackett was shutting down, Garry Lucas left the flightdeck and opened the freight door. The nose landing gear having collapsed, the doorway sill was only a short distance above the ground so the first officer was able to get quickly outside and encourage the passengers in their escape. All emergency lighting functioned normally.

Meanwhile Helen Dutton checked for fire outside the left rear door and then opened it. Because the fire was still burning fiercely and smoke was starting to enter the cabin, it was thought quickest to direct the passengers to jump directly from the doorway, which was some 11ft above the ground.

It is estimated that the evacuation was accomplished in less than a minute, with 16 passengers escaping through the rear exit, a further 16 through the over-wing exit and the remaining eight through the freight doorway. The first officer and second cabin attendant also left through the freight doorway, as did the commander after he had assured himself that there was no one left aboard. Once the last passenger had left through the rear exit, Helen Dutton moved forward through the cabin and carried out a thorough check of all the seats before returning to the rear door and leaving the aircraft.

The accident inquiry concluded that the crew's actions in safely evacuating all 40 passengers from the Emerald Airways flight without fully completing the evacuation drills were 'understandable in view of the severity of the fire'. But they were lucky to get away with it. Passengers who moved to the rear exit found that the escape chute could not be deployed because they were standing on the stowage hatch. Whoever designed G-OJEM's escape system did not seem to have registered that this would be a likely happening in an emergency evacuation situation.

Even if the stowage hatch had been accessible, it seems likely that chute deployment would have taken an inordinate length of time. An automatically inflated emergency slide system had once been installed on G-OJEM, but sometime in the past it had been taken out and replaced with a manual variant. The manual emergency escape procedure was to find and open the stowage hatch, attach two side panel clips, throw out the chute, brief and dispatch two 'Able Bodied Persons' down the rope and wait for them to find the handles and tension the chute. This would have had to be carried out at night, with only emergency lighting, and with the added stress of a substantial engine bay fire burning outside the windows.

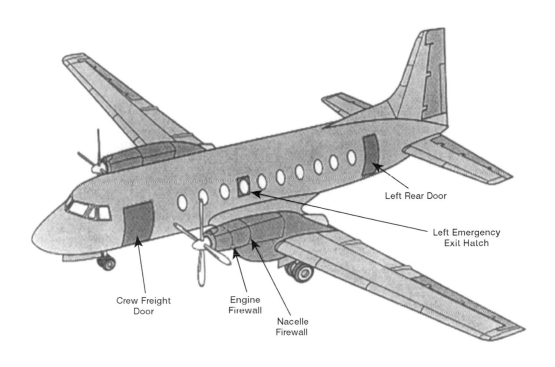

Left Rear Door

Left Emergency
Exit Hatch

Crew Freight
Door

Engine
Firewall

Nacelle
Firewall

The height of the rear HS748 exit sills is relatively modest when the aircraft stands on all three undercarriage legs, but it increased considerably once the nose gear collapsed and the tail went up. It seems likely that the evacuation time on 30 March 1998 was appreciably shortened by the decision to evacuate the aircraft without attempting to deploy the rear chute. Fortunately, no major injury resulted, although the drop to the ground was appreciable. Provision of an effective emergency escape system, usable without injury by passengers of all ages and levels of physical agility, is not an unreasonable expectation on a commercial airliner.

Among the 19 safety recommendations made after the investigation into what could have been a very nasty accident at Stansted, the British Civil Aviation Authority required the HS748 to be provided with an easily deployable and effective means of descent from both fuselage rear doors in an emergency evacuation situation with any landing gear configuration. The aircraft's engine manufacturer was also asked to reassess the susceptibility of the three-stage Dart turbine to failure and to ensure that effective action aimed at preventing recurrence had been taken. The CAA also sought modification of the PA/interphone system on HS748 aircraft at the rear cabin attendant position to make the handset and its controls accessible to an attendant strapped into the aft cabin crew seat. It was not very reassuring to the paying public that those responsible for their safety in the air had taken nearly 40 years to come up with that.

Above:
General layout of the port side of the HS748.

Finally, what of the pilots' handling of the emergency? All HS748 pilots will have been taught that the classic engine failure/fire response at or after V1 is for the handling pilot to continue the take-off and, when safely established in the climb, for the non-handling pilot to action the immediate Engine Fire/Failure Check List from memory. The aircraft would then be positioned for an expeditious approach back at the departure airfield or to a suitable alternate.

The sequence of events on 30 March did not match this classic scenario. The first indication the pilots had that anything was amiss came shortly after the aircraft got airborne when they heard a noise, identified in the course of the investigation as being made by the HP turbine disc leaving the engine. The sparks coming from the No 2 engine jet pipe, which were the first warning of the abrupt failure and uncontained release of the errant HP turbine, were not visible to the flight crew. It took audible shouts from the passengers to make the pilots aware of a substantial fire on the starboard wing area around the engine. It was at about the time that Helen Dutton reported the right engine on fire that the commander took control. Fearing for the structural integrity of the aircraft, and aware that a considerable stretch of the 3,048m runway remained ahead, Capt Hackett made the decision to re-land and reduced power on the left engine. Once he made that decision, it was irrevocable. It was subsequently that the engine fire warning bell sounded, some 12 seconds after the engine failure.

Although yaw was increased momentarily by the water methanol system kicking in on the live engine, the commander managed to place the HS748 back on the runway with about 448m of paved surface remaining. Notwithstanding whatever is written in Operating Manuals, an aircraft commander has the ultimate discretion to act as he deems best in an emergency. The 61-year-old Capt Hackett oozed experience and airmanship, with 6,100 flying hours in his logbook, of which 3,950 were on HS748s. Given that the first officer had only 250 hours on the type, Capt Hackett's decision to take control from the competent but far less experienced first officer and to land back on, based on the limited information available to him at that instant, was sensible in the circumstances. It would been impossible for the commander to have communicated his decision to re-land to First Officer Lucas quickly enough to allow him to fly the demanding handling manoeuvre in the time available.

The commander of what became known as 'the Leeds United aircraft' got the backing of air investigators for his unorthodox but successful handling of his HS748 when it crashed. The AAIB and media verdict might have been very different if G-OJEM had burst into flames as it left the runway, killing a host of passengers including a very expensive soccer team. But Capt Hackett hacked it, and that was what mattered. The moral for the future is that even if an aircraft has been in service for upwards of four decades, it can still bite. And everyone, be they passengers or crew, should stay alert for any eventuality until their aircraft is safely established in the climb.

3

Conspiracy Theories

Back in 1959, Britain's most modern strategic jet bomber was the Handley Page Victor Mk 2. Derived from the Victor Mk 1 by the classic 'stretching' process of bigger engines, more wing area and higher all-up weight, the first Victor 2 (XH668) was designed to operate at up to 60,000ft to get above the first generation of Soviet surface-to-air missiles then entering service.

After manufacturer's proving flying, XH668 was transferred to the Boscombe Down experimental and evaluation establishment for preview handling trials. The first trials flight, scheduled for Wednesday 20 August, called for a climb to 52,000ft, an hour's tests including high-speed turns to reach the fringe of wing buffeting and to pass a little beyond it, a rapid descent using airbrakes down to 35,000ft, and a further series of tests before returning to Boscombe after a trip lasting 2-3 hours.

XH668 got airborne at 10.35hrs local time. A warning of an aircraft passing ahead was acknowledged by XH668 at 10.38hrs, but Boscombe had no further radio contact with the Victor crew. Although a Boscombe Flying Order stated that prototype aircraft should maintain communications with the ground, XH668 was regarded as a new mark of an existing type so it was not considered to be a prototype under the terms of this Order. Furthermore, because of weather vagaries and the large distances covered at high speed, pre-flight submission of a precise flight plan was not expected and therefore Boscombe had only a rough idea of when and where the crew intended to operate. Whatever the rights and wrongs of this loose-leash system, no one was officially keeping an eye on XH668. It was only because the Victor captain had asked for his landing run to be photographed that anyone noticed anything amiss as early as they did. At about 13.10hrs the photographer made a telephone call to ATC asking for XH668's estimated time of arrival. After some confusion because nobody was exactly sure how long the test crew intended to stay airborne, 'overdue' action was taken at 15.03hrs. By then, XH668 had long since disappeared from the sky.

The disappearance of Britain's latest bomber conjured up all manner of theories, including hijacking along the lines of James Bond's *Thunderball*. Many people said that they had seen an unfamiliar shape 'flying at a great height and very fast', but most sightings were either too late in the afternoon or in unlikely places, such as 'low flying over Kensington'. There were also the usual weird offerings, such as that from the Parisian gentleman who attributed the loss of XH668 to mysterious but none the less damaging powers unleashed when the Moon rose and set.

On a small coaster bound from the Mersey to the Thames on 20 August, the master heard a BBC radio broadcast that an aircraft was overdue from Boscombe Down. Earlier that day, when the coaster was off St David's Head, Pembrokeshire, the master and two of his crew had been on the bridge because there were extensive fog patches about. At around 11.40hrs the vessel ran out of the fog and the seamen observed a large column of water and spray about 50ft high and some five miles away, followed almost immediately by two sharp reports similar to rifle fire. Fortuitously, radar stations kept films of all responses seen on their screens and RAF Wartling in Sussex found a radar track that ended abruptly at approximately the same position and time as that reported from the coaster.

The final radar response from the Victor showed it to be in a turn to port and the corrected position of this last response and the position of the splash were found to be only 10 miles apart. The view that at least a substantial portion of XH668 entered the water at the reported position was confirmed on 25 August when the first white fragment of Victor radome was found by a schoolboy on White Sands beach, St David's.

Why did the Victor go down when it did? The weather had been good at the time, and there were no signs of clear air turbulence or jet streams to trouble the crew, whose members were more than capable of carrying out the task they had been set. There was an adequate supply of oxygen on board the bomber and as individual crewmembers were wearing air-ventilated suits, anti-g trousers, partial-pressure jerkins and helmets, they should have been protected from the physio-logical effects of high flight. No single fault could cause total loss of electric power on the Victor 2, and the entire flight was planned to remain within design limitations. There should have been no fuel tank or battery explosion because the Victor was flying with unpressurised fuel tanks and a low-voltage battery that was not crucial to the maintenance of all electrical power.

It was possible that the bomber had been sabotaged, but Boscombe was a very secure airfield patrolled at night by police with dogs, and to wipe a Victor off the screen at one fell swoop would have required precise knowledge about take-off time, duration of flight and the vulnerable parts of the aircraft's structure. How-ever, the plotted track of the missing Victor passed close to the missile range at Aberporth, which around that period was testing the Bloodhound surface-to-air missile, designed specifically to cope with high-flying jet bombers. Suppose that a Bloodhound had been test-fired out over the Irish Sea and the Victor was heading in that direction. The Bloodhound suddenly went astray, its controller found that his destruct button would not function, and the latest British air defender collided with the latest British attacker. It must have been with a sigh of relief that the inquiry found that there had been no missile launches from Aberporth on 20 August, nor was there any simultaneous activity in adjacent danger areas where interceptors may have been firing guns or air-to-air missiles.

But the only way to make sure that nothing had hit the bomber was to check the wreckage, and unfortunately that was 400ft down in the bleakest and roughest

patch of sea anywhere in the vicinity of the British Isles. XH668 weighed around 63 tons when it crashed, of which 20 tons was fuel. Over the next 15 months, 1,480 men and 40 vessels eventually retrieved 592,610 pieces of Victor. The plan was for XH668 to be more or less rebuilt from its broken remains, and in fitting the jigsaw back together it was hoped to discover why it broke up in the first place.

Mounted on each Victor 2 wingtip was a long, slender tube projecting 6ft ahead of the leading edge comprising both static and pitot pressure systems. When the wingtips were reassembled, the starboard wingtip pitot tube was missing in its entirety. A Victor pitot tube was held in position by a form of chuck in which a conical sleeve was tightened against two tapered collets. There was nothing wrong with the design but the starboard pitot tube had come cleanly and neatly out of its mounting without any signs of damage. Furthermore, when the tapered sleeve for the starboard mounting was recovered from the sea with the collets still in position, the inside of the sleeve was found to be covered in protective paint which had clearly been applied too freely. It was surmised that the sleeve had originally been tightened adequately but only against localised areas of paint. As the paint wore away or became distorted, progressive

Above:
Prototype Victor Mk 2 XH668 at Handley Page's Hertfordshire factory on 13 March 1959. One of the Victor's pitot tubes, which caused all the trouble on 20 August 1959, is visible jutting out from the port wingtip.
Author's collection

loosening of the collets' grip upon the pitot tube could reasonably be expected. Nothing similar was found on the port installation.

The starboard tube was unique in serving the 'stall detector' and the Mach trimmer. Wind tunnel tests a decade earlier had predicted that a Victor crescent wing stall would be pretty vicious, so Handley Page developed great accumulators of stored energy which thumped leading edge flaps down within a second on a signal from a pressure ratio switch which calculated lift coefficient from tappings below the wings. While the flaps travelled, warning lights illuminated in front of both pilots. In addition to pressure ratio switch signals, the stall detector received pitot and static pressures from the starboard pitot tube.

The Mach trimmer counteracted the nosedown trim change caused by air compressibility at a high Mach number. As the aircraft accelerated from Mach 0.85 to Mach 0.95, the Mach trimmer raised the elevators without altering the stick position, irrespective of the pilot's control inputs, in response to pressures conveyed from the starboard pitot tube.

Investigators assumed that XH668 was flying buffeting turns around Mach 0.94 at 52,000ft. Suppose then that the starboard pitot pressure tube started to come adrift slowly under the effect of all the buffeting. A leak would slowly develop causing a fall in indicated airspeed on the co-pilot's side. Assuming this was noticed, the captain's instruments would still be functioning properly and maybe the assumption was made that a bit of water in the starboard system had frozen up.

So far so good, but then the pitot tube dropped clean away. Zero speed would have been recorded in front of the co-pilot but as the tube fell away, the stall detector would have lowered the nose flaps. The first the pilots might have known about this was when the warning lights came on, but what would they have made of them? At Mach 0.94 and 52,000ft, XH668 would have been bucking and rearing round the stops as its crew explored the steep turn buffet boundary, and in such circumstances the warning lights and lowering nose flaps would only have confused matters because they would have been activated for no reason as far as the crew were concerned.

Faced with stall warning indications, the natural reaction would have been to lower the Victor's nose. Unfortunately, these distractions would have masked the most sinister 'gotcha'. At Mach 0.95 the Mach trimmer would have been fully out, but on receipt of a spurious low-speed signal it would have steadily lowered the elevators and pushed down the aircraft's nose which was very efficient aerodynamically. Moreover, the elevators would be lowering at a time when the pilot might have been pushing the stick forward. The combined nosedown movement — and it might have been increased if the co-pilot had instinctively reacted to an apparent fall in his airspeed indicator before realising what was happening — would have pushed the Victor quickly beyond human recovery. It is most likely that the throttles stayed open and the airbrakes closed because both pilots were concentrating all their hands and efforts on trying to overcome the forces opposing the aileron jacks to level the wings as an essential preliminary to recovering from the dive. It would have been a futile effort and the crew were doomed as the spiral tightened. On deciding that the wings were now at a low incidence, the stall detector should have sent a signal to raise the nose flaps, but the aerodynamic forces experienced in the high-speed dive would have resisted their operating jacks as well.

XH668's nose flaps were found to be down and its Mach trimmer actuator virtually fully retracted when it hit the sea. The jigsaw fitted together to provide all the ingredients of a major disaster. To prevent recurrence, the mountings of the pitot tubes of all Victor aircraft were modified such that the collets were locked permanently and therefore could not vibrate loose in future under any circumstances. As the Victor stall was nothing like as bad as feared, the leading edge flaps were permanently locked up.

Right at the bottom of St George's Channel, a Victor 2 pitot head lies buried in the sand and is likely ever to remain so. The saga of the lost Victor 2 prototype reinforced the point that all accident causes must be determined satisfactorily if faith is to be retained in an aircraft or its operating procedures. The only firm record of what went on inside XH668 that fateful day was a fragment of the crew's last conversation transmitted in error and picked up purely by chance at Boscombe. It was a very weak and mangled transmission and by a freak of radio it told the inquiry that the crew were listening to a daily soap entitled *Mrs Dale's Diary* as the aircraft went down.

The future lay in developing flight and cockpit data recorders. XH668 carried a primitive wire recorder during its test flight and some 2,500ft of wire was recovered in what first appeared to be a hopelessly tangled and broken mass. After hundreds of man-hours had been spent disentangling and joining pieces together, it was established that the wire came from an unused spool. Technology has progressed apace since 1959: modern data recorders not only record voice inputs from the crew as well as other noise from area microphones but also take down crucial parameters such as height, speed, acceleration forces, control surface angles, pitch, roll and yaw information and engine power output. Data recorders, capable of being easily 'milked' by computers, are now a must, and it goes without saying that such 'black boxes' — they are in fact painted yellow to aid recovery — have to be crash-protected.

When a large aircraft crashes in dramatic circumstances, the tendency is to look for a big, headline-grabbing cause rather than a small, mundane one. Yet far too many flying accidents are still caused by silly little things such as a loose spanner jamming a control run, a carelessly dropped bulldog clip sucked into an engine intake or a decimal point put in the wrong place when calculating turbine blade clearances. If the loss of XH668 taught anything, it was that everybody involved in building, operating or maintaining aircraft, no matter how remote from the flight line the task, must always give it their utmost care. Quality control and effective supervision are not just high-tech preserves; they apply right across to the chap with the paintbrush.

Thirty-seven years later, Boeing 747-131 N93119 belonging to Trans World Airlines departed Athens, Greece, just after dawn on 17 July 1996. It landed at John F. Kennedy International Airport (JFK), New York, around 16.31hrs Eastern Daylight Time, and shut down at Terminal 5, Gate 27, some seven minutes later. No operational abnormalities were noted during the flight. A new crew took over N93119. It was refuelled and remained at Gate 27 for about 2½ hours until the time came to depart as TWA Flight 800.

Up front, a 58-year-old captain with 18,800 flying hours occupied the left front seat, a 57-year-old captain/check airman with 17,000 flying hours occupied

the right front seat, a 24-year-old flight engineer occupied the right aft seat, and a 62-year-old flight engineer/check airman occupied the left aft cockpit jump seat. Flight 800 was scheduled to depart for Paris at about 19.00hrs, but pushback was delayed because of a disabled piece of ground equipment and concerns about a suspected passenger/baggage mismatch. According to the cockpit voice recorder (CVR), gate agent personnel advised the flight crew just before 20.00hrs that although a passenger's bag had been pulled, the passenger had been on board the whole time. The CVR recorded the sound of the cockpit door closing at 19.59:59, and the 747 was pushed back from the gate about 20.02. The flight crew started Nos 1, 2 and 4 engines between 20.05 and 20.07:46. At 20.07:52, the captain/ check airman advised the JFK gate hold controller that TWA800 was ready to taxi. About 20.08, the 747 began to taxi towards departure runway 22 Right. While the airliner was taxying, the flight crew started No 3 engine and completed delayed engine-start and taxi checklists.

N93119 got airborne at about 20.19. During departure from JFK, the pilots received a series of generally increasing altitude assignments and heading changes from New York Approach and Boston ARTCC controllers. Flight 800 reached 13,000ft at 20.27:47. At 20.29:15 the CVR recorded the captain stating, 'Look at that crazy fuel flow indicator, there on number four . . . see that' A minute later, Boston Center advised TWA800 to climb and maintain 15,000ft. The captain in the left front seat, who was flying the 747, asked for 'climb thrust'. The captain/check airman acknowledged the ATC clearance at 20.30:18 and seven seconds later, the captain repeated, 'Climb thrust'. The flight engineer, who was receiving initial operating training, responded, 'Power set'.

The airliner was climbing through 13,760ft at 20.31:12 when both the CVR and the flight data recorder ceased recording without warning. At 20.31:50, the captain of an Eastwind Airlines Boeing 737 (Stinger Bee Flight 507), about 35km northeast of Flight 800 on a southwesterly heading, told Boston Center that he 'just saw an explosion out here'. About 10 seconds later, Stinger Bee 507 transmitted, 'We just saw an explosion up ahead of us here — about 16,000ft or something like that, it just went down into the water.' ATC facilities in the New York/Long Island area subsequently received reports of an explosion from other pilots operating in the area.

Many witnesses in the vicinity stated that they saw and/or heard explosions, accompanied by a large fireball over the ocean, and observed debris, some of which was burning, falling to the water. About one-third of these witnesses reported that they observed a streak of light, resembling a flare, moving upward in the sky to the point where a large fireball appeared. Several witnesses reported seeing this fireball split into two as it descended toward the water. What everyone was witnessing around 20.31hrs was TWA Flight 800 crashing into the Atlantic Ocean about eight miles south of East Moriches, New York. All 230 people on board — two pilots, two flight engineers, 14 flight attendants and 212 passengers — were killed, and the 747, valued at $11 million, was destroyed.

*   *   *   *   *

At the time of the accident, there were light winds and scattered clouds in the area, but there were no significant meteorological factors that might have disrupted the flight. The flight crew, who had 93 years of TWA time between them, were properly certificated and qualified, and in receipt of the training and off-duty time prescribed by Federal regulations. There was no evidence of any pre-existing medical or behavioural conditions that might have adversely affected the flight crew's performance during the fateful flight.

The witnesses' reports, plus the widespread distribution of the wreckage, indicated that Flight 800 had experienced a catastrophic in-flight structural breakup. This view was reinforced by the sound in the last few tenths of a second before the CVR recording stopped which was similar to the last noises heard on CVR recordings from other aircraft that had experienced structural breakups. Consequently, investigators considered the following possible causes for TWA800's in-flight breakup: structural failure and decompression; detonation of a high-energy explosive device, such as a bomb exploding inside the 747 or a missile warhead exploding upon impact with it; and a fuel/air explosion in the centre wing fuel tank.

Pieces of Flight 800 wreckage were distributed along a northeasterly path about 4 miles long by 3½ miles wide in the Atlantic Ocean off Long Island.

Remote-operated vehicles, side-scan sonar and laser line-scanning equipment were used to search underwater, and scuba divers and robot vehicles were used to recover victims and wreckage. In the later stages of

Above left:
N93119, the Boeing 747 which went down as TWA Flight 800.
The Aviation Picture Library

wreckage recovery, scallop trawlers were used to recover pieces of wreckage that had become embedded in the ocean floor.

These pieces of wreckage were tagged and numbered according to their recovery location. Almost all pieces of wreckage were then transported by boat to shore through the Shinnecock Inlet, where they were loaded onto trucks and transported to leased hangar space at the former Grumman Aircraft facility in Calverton, New York. On arrival at the hangar, investigators set about trying to identify and document the pieces of wreckage.

The recovery effort took more than 10 months and involved many agencies and companies. Most of the examination, documentation and pertinent reconstruction of the recovered pieces of wreckage was completed within one year of the accident. Some wreckage examination was ongoing until mid-2000, and it was a tribute to many people and much arduous work in difficult conditions that more than 95% of the wreckage of N93119 was eventually recovered.

N93119 was a 747-131 series airliner manufactured by Boeing in July 1971 and purchased new by TWA. It had amassed 93,303 total hours of operation (16,869 complete take-off and landing sequences) at the time of its loss, but despite the 747's age, close examination of the wreckage revealed no evidence of structural faults (such as fatigue, corrosion or mechanical damage) that could have contributed to the in-flight breakup. Breakup could have been initiated by in-flight separation of the forward cargo door, but all the evidence indicated that the door was closed and locked at impact. Investigators concluded that the in-flight breakup of TWA800 was not initiated by a pre-existing condition resulting in a structural failure and decompression.

Of the 736 witnesses to TWA 800 in its dying throes, 258 described seeing a streak of light, a flare-like object, or fireworks in the sky about the time of the

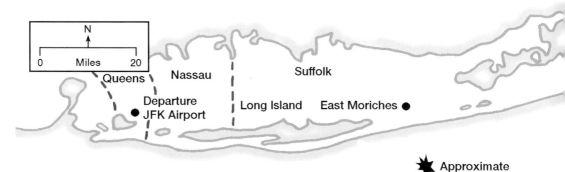

accident. There was intense public interest in these witness reports and much speculation that the reported streak of light was a missile that eventually struck TWA800, causing the 747 to explode.

It was not unreasonable to speculate that the accident might have been caused by a bomb explosion or missile strike. The 1996 Olympic Games were then being held in the United States, and the sudden and catastrophic in-flight breakup of TWA800 was undeniable. There were certainly some unusual primary radar targets recorded by the radar site at Islip, New York, in the general vicinity of TWA800 at the time of the accident. Investigators closely evaluated three sequences of Islip radar returns from about one minute before the accident to nine minutes afterwards. These appeared to show three targets moving at 300-400kt groundspeed about 10nm southwest of the accident. However, none of the three sequences intersected TWA Flight 800's position at any time, and all of them were moving away from the 747. Investigators concluded that the three Islip radar sequences recorded between about 20.30 and 20.40 did not represent an unexplained object such as a missile but, rather, they were false or reflected returns from actual aircraft in other geographic areas.

None the less, in view of the theories abounding in the media, the Safety Board considered the possibility that a bomb exploded inside N93119 or that a warhead from a shoulder-launched missile blew up the airliner. Throughout the wreckage recovery and evaluation processes, fire, explosives and metallurgy experts from the NTSB, Department of Defense, FBI, Bureau of Alcohol, Tobacco & Firearms and the Federal Aviation Administration thoroughly examined all recovered pieces of the wreckage for evidence of damage that may have been caused by a bomb, high explosive or missile warhead entry or detonation. No evidence of any such damage, which is pretty distinctive, was found. Furthermore, victims' remains showed no evidence of injuries that could have been caused by high-energy explosives, nor was there any damage to the 747 seats or other interior components consistent with a high-energy explosion.

Trace amounts of explosive residue were detected on wreckage pieces from TWA800, but testing by Federal Aviation Administration technicians showed that residues of explosives would have dissipated completely after two days of immersion in seawater. Very few pieces of aircraft wreckage were recovered during the first 48 hours after the accident, so it was deduced that the explosive residues were not present on N93119 when it entered the water. It was decided that the explosive particles were deposited during or after recovery operations from the military personnel, ships or ground vehicles used during the recovery operation. This was a very convenient explanation, but the important fact to bear in mind is that the lack of any corroborating evidence associated with a high-energy explosion indicated that these trace amounts did not result from the detonation of a high-energy explosive device on Flight 800. Unless someone had launched a revolutionary new weapon at TWA800, the effects of which had never been seen before, it was safe to say that the in-flight breakup of N93119 was not initiated by a bomb or a missile strike.

But something blew TWA800 apart. Wreckage found closest to JFK along the 747's flight path, and

Left:
The TWA800 crash site.

therefore the earliest pieces to depart the airliner, consisted primarily of debris from the 747's wing centre section, which included the centre wing fuel tank (CWT). From pieces of wreckage lightly covered in soot, investigators concluded that the start of the breakup sequence was an over-pressure event within the CWT. And because there was no evidence that a high-energy explosive device detonated in this (or any other) area of the aircraft, this over-pressure could have been caused only by a fuel/air explosion in the CWT.

In bald terms, an explosion in the CWT blew the nose off the 747. Computer simulations based on radar data, trajectory calculations and aircraft performance factors indicated that the remainder of the airliner continued in crippled flight and pitched up while rolling to the left (north). In the process the crippled 747 ascended from 13,800ft to about 15,000-16,000ft, before rolling into a descending turn to the right (south). The time interval between the CWT explosion and the aft portion of the 747 hitting the water was 47-54 seconds. It is likely that, after the nose portion separated from the aft fuselage, a fuel-fed fire within the breached CWT (or any other fire that might have existed, such as from fuel that might have been leaking at the wing roots) would have been very visible and was probably the streak of light reported by many witnesses.

According to TWA's dispatch documents, N93119 took off with 176,600lb of fuel, but it was a tribute to the range of the 747 that the 13,000-gallon centre wing fuel tank could be left empty. Only 50 gallons was slopping around the bottom of the CWT as N93119 passed 13,000ft, which meant that there was a lot of room for volatile vapour. This would normally vent off but investigators assessed that fuel/air vapour in the ullage of TWA800's CWT was flammable at the time of the accident. A fuel/air explosion in the centre wing fuel tank at this stage would have generated sufficient internal pressure to break apart the tank. What was less clear was what had initiated the explosion of the vapour.

No lightning was in the area at the time of the accident and no report existed of a meteorite ever having struck an aircraft. Not wanting to be accused of any oversight, investigators considered whether a missile might have self-destructed close enough to TWA800 for a fragment to have entered the CWT and ignited the fuel/air vapour, while being far enough away not to leave any damage characteristic of a missile strike. The Safety Board calculated that such a detonation would have to occur within 40ft of an aircraft for a fragment travelling perpendicular to the CWT to retain sufficient velocity to penetrate the tank. In the case of a fragment travelling at an angle to the CWT, a detonation would have to occur even closer to the aircraft. Given that none of the holes in the CWT or fuselage structure exhibited characteristics of high-velocity penetration that would have resulted from a missile fragment, it seemed very unlikely that a missile warhead self-destructed at a distance close enough to force at least one fragment into the CWT with sufficient velocity to ignite the fuel/air vapour therein, while being far enough from the 747 for none of the remaining fragments to impact with its structure.

Detonation of a small explosive charge placed on the CWT could have breached the fuel tank and ignited the flammable fuel/air vapour. However, testing by the British Defence Evaluation & Research Agency in 1997 had demonstrated that when metal of the same type and thickness as the CWT walls

was penetrated by a small charge, there was 'petalling' of the surface on which the charge was placed, pitting on the adjacent surfaces, and visible hot gas washing damage in the surrounding area. As none of these damage characteristics was found on the recovered CWT wreckage, it was very unlikely that a small explosive charge detonated on the CWT, breached the fuel tank, and ignited the fuel/air vapour.

Electromagnetic interference from personal electronic devices was found to have played no role in igniting the vapour, so investigators then evaluated whether the surface of the CWT might have been heated to a high enough temperature to have caused ignition of the fuel/air vapour inside the tank. A bleed air leak near the CWT, a fire in the air conditioning pack bay beneath the CWT or a fire in the main landing gear wheel well could have raised a large volume of the CWT ullage above the auto ignition temperature or caused a localised hot spot on the interior of the CWT skin sufficient to reach hot surface ignition temperatures.

However, a scenario in which a significant portion of the CWT volume was heated to auto ignition temperatures was not considered a credible possibility because the amount of heat that would have been necessary to raise the temperature of most of the CWT to at least 460°F would have almost certainly left significant evidence of thermal damage on the CWT. These

Above:
The watery grave of TWA800 floating off East Moriches, New York.
Associated Press

temperatures would be near the softening and melting temperatures of the CWT's aluminium skin, but no signs of this were found. The engines, all of which were recovered and extensively examined, showed no evidence of any failure or malfunction. Furthermore, engineering data from the manufacturer of the air conditioning system and Boeing indicated that the combination of failures required to produce such elevated bleed air temperatures would also have caused an overpressure leading to an air conditioning duct burst. The reconstructed air conditioning packs showed no evidence of leakage in the packs or ducts. In addition, no discussion of an overheat or other warning was recorded by the CVR, which would be expected if excessive bleed air temperatures or any other failure occurred, especially with two flight engineers in the cockpit.

Investigators concluded that it was very unlikely that the flammable fuel/air vapour in the CWT was ignited by high temperatures produced by sources external to the CWT; a fire migrating to the CWT from another fuel tank via the vent system; an uncontained engine failure or a turbine burst in the air conditioning packs beneath the CWT; a malfunctioning CWT jettison/override pump; a malfunctioning CWT scavenge pump; or static electricity.

Which left a good old-fashioned short circuit. There were several pointers to possible anomalous electrical events just before the explosion. First, a buzz on the captain's CVR went silent less than a second before the CVR lost power. Second, the captain's comments about a 'crazy' No 4 (right outboard engine) fuel flow indicator, recorded on the CVR some two minutes before it lost power, also suggested some type of electrical anomaly affecting that wiring. There had also been recent problems with lights on the 747.

Boeing design specifications permitted fuel gauge indication wiring to be bundled with or routed next to higher-voltage aircraft system wires, some carrying as much as 350 volts. Excess voltage from a short circuit can be transferred from wires carrying higher voltage to wires carrying lower voltage if the wires are near each other. Evidence of arcing was found on generator cables routed with wires in the leading edge of the right wing, near the wing root. Because this wire bundle included wires leading to the right main wing tank (No 4) fuel flow gauge routed to a connection in the CWT, a short circuit in this bundle could have carried excessive energy into the CWT fuel indication system.

Consequently, NTSB investigators concluded that TWA Flight 800 was lost because of an exploding fuel/air mixture in the centre wing fuel tank. The ignition source for the explosion could not be determined with certainty, but the most likely cause was a short circuit outside of the CWT between the low-voltage wire used by the fuel gauges and high-voltage wires used for other aircraft systems. Contributing factors to the accident were the design and certification concept that fuel tank explosions could be prevented solely by precluding all ignition sources, and the fact that the Boeing 747 had been designed with no means of reducing heat transferred into the CWT or to render fuel vapour in the tank non-flammable.

Subsequently, the Federal Aviation Administration studied ways to reduce volatility by infusing fuel tanks with inert nitrogen gas, a safety precaution that military aircraft had used for decades. The NTSB also urged national aviation authorities to examine manufacturers' design practices to eliminate potential ignition hazards involving fuel tank components, to review aircraft design specifications to make sure low-voltage fuel tank wires were separated from high-voltage wires, and to require companies to improve training of maintenance personnel so that they would recognise and repair wiring problems in ageing aircraft.

What was the upshot of the loss of TWA800? At the personal level, the 21 families from Montoursville, Pennsylvania who lost spouses or children on TWA800 settled with TWA and Boeing for more than $2 million each. Boeing also agreed to give $100,000 to help build a memorial on Long Island for the 230 people killed in the crash.

Given the unprecedented media and political interest in the accident, the US Government left no stone unturned to try to determine what had happened to N93119. During the 17-month, $40million Federal investigation, the US Coast Guard trawled for wreckage in 120ft-deep waters, over 500 FBI agents interviewed more than 7,000 people, and investigators from the NTSB, FAA, State Department and numerous local and state law enforcement agencies complemented the engineers who painstakingly reconstructed 95% of the big jumbo jet. Like the Victor investigation four decades earlier, it was the biggest overt aircraft recovery exercise to date.

James Kallstrom, who led the FBI investigation, reported that 'every lead has been covered, all possible avenues of investigation exhaustively explored and every resource of the United States Government has been brought to bear in this investigation . . . We must now report that no evidence has been found which would indicate that a criminal act was the cause.'

The CIA went so far as to commission a computer animation, showing how a centre wing tank explosion ripped the front end off the 747. This was packaged within a video movie of the accident investigation, in the hope that such openness would dispel claims that the Government botched the investigation or was engaged in a cover-up.

The fact that at least 258 witnesses saw streaks of light in the dusk as TWA800 fell out of the sky fuelled speculation that a missile may have brought down the 747. However, analysis of these observations did not accord with a missile attack. Rather, they were consistent with observing some part of the in-flight fire and breakup sequence after the centre wing tank explosion. 'We firmly believe the witnesses told us what they saw,' said Kallstrom. But because sound travels more slowly than light, when witnesses heard the sound of the blast and thought they were seeing the start of the crash, they were in fact watching its end.

NTSB Chairman Jim Hall was to state that, 'It is unfortunate that a small number of people, pursuing their own agendas, have persisted in making unfounded charges of a Government cover-up in this investigation. These people do a grievous injustice to the many dedicated individuals, civilian and military, who have been involved in this investigation.' Suffice it to say that too many professional people were involved in investigating the loss of TWA800 for any cover-up to have remained covered up this long.

'I would love to walk in here with a molten piece of wire and say, "Here it is",' said senior accident investigator Robert Swaim. In the absence of a smoking gun, the FAA reviewed the history of fuel tanks on 10,000 commercial and non-combat military aircraft. Airliners were studied during their regular maintenance stops, including those belonging to other carriers and other countries. Inspectors looked specifically for wiring problems, and although inspections on 81 airliners found 'room for improvement', there were no immediate safety problems.

But the FAA found definite room for improvement in maintenance practices. Inspectors discovered damaged wiring insulation, improper and frequent repairs of wires, the presence of metal drill shavings in wire bundles, opened splices that should have been sealed and lint on circuit breakers. In some cases, wiring had been damaged by anti-corrosion spray and overflow from lavatories.

Although investigators never believed that the problem reached a level where airliners were unsafe, the FAA ordered 37 corrective actions including replacing sharp-edged fuel probes that might damage wires, keeping pumps idle unless they are submerged in fuel, installing protective sleeves on wiring in tanks, and developing electronic devices to suppress power surges in wiring.

A three-year FAA study eventually concluded that airline fuel tanks were safe, but they had to be kept that way. As not all wiring problems were caused by ageing — in some cases, wiring insulation had been unintentionally damaged by work crews in tight areas — Boeing and Airbus advised airlines of problem areas so that they could improve wiring inspection procedures and documentation to show fully repairs made on wiring. The FAA improved its training of inspectors better to focus on wiring problems, and it is working on the development of new, more sensitive circuit breakers to shut off power when a short occurs, and on technology to check the condition of wires throughout an aircraft. Finally, serious attempts are being made to overturn the decades-old philosophy of minimising the risk of explosion by reducing possible ignition sources rather than reducing the volatility of fuel tank contents. The loss of TWA800 showed that 'all possible ignition sources cannot be predicted and reliably eliminated'.

What did the TWA flight crew think of the NTSB judgement? It has to be said that the NTSB verdict did not initially wash with most TWA 747 pilots who, since 1970, had been operating with similar procedures and in near identical atmospheric and operating conditions. The unacknowledged friendly fire scenario is still entertained by some, although it is difficult to imagine it being concealed by all those involved for so long. Eventually many 747 pilots came to accept the NTSB's probable cause, although for quite a few the jury is definitely still out.

Jim Majer, a former TWA captain whose flying career stretched back 40-plus years, was convinced that the flight crew were blameless. 'I knew and had flown with all the pilots, both operating and deadheading, with the exception of the student flight engineer. I had previously flown with most of the operating and deadheading flight attendants. The check pilot had previously qualified me on the Gulf Air L-1011 and given me various check rides through the years. The flight engineer checker and retired captain Dick Campbell and I had flown together since 1965 on Constellations, 707s and 747s. The captain being checked was highly regarded by his peers. There is not the slightest doubt in my mind that company procedures were followed exactly by the book and to the letter.'

Jim spoke for many flight crew when he said that 'Personally, I have come to accept NTSB's "probable" cause, although the spectre of an unlikely friendly fire cover-up haunts me as a remote possibility. The technology does exist, so it therefore cannot be completely ruled out.'

In spite of all the efforts to make airline fleets much safer, the TWA800 saga will rumble on and on because the NTSB never came up with a solid body of firm evidence to close down the conspiracy theorists. After outlining the exploding centre wing tank theory in great detail, the NTSB Director of Aviation Safety, Bernard Loeb, conceded that 'We cannot be certain that this, in fact, occurred, but of all the ignition scenarios we considered, this is most likely.' Notwithstanding all the advances in aircraft design, technology and aviation wisdom since XH668 went down off Pembrokeshire, the investigation of high-profile aircraft accidents under an ever more piercing media spotlight is not getting any easier.

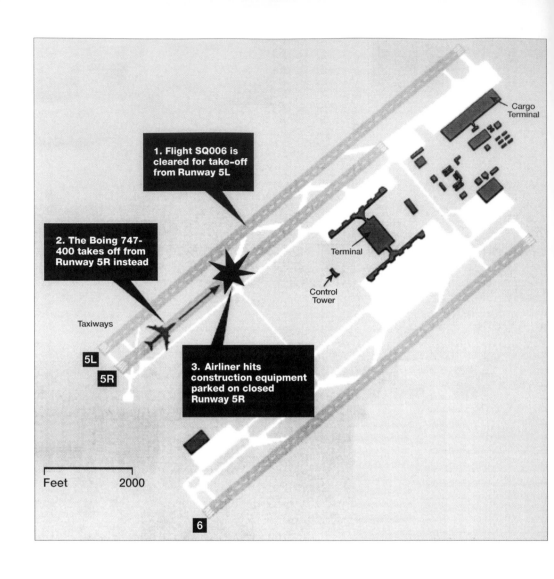

1. Flight SQ006 is cleared for take-off from Runway 5L

2. The Boing 747-400 takes off from Runway 5R instead

3. Airliner hits construction equipment parked on closed Runway 5R

Cargo Terminal

Terminal

Control Tower

Taxiways

5L

5R

Feet   2000

6

Above:
Pictorial representation of the take-off roll by SQ006 on Runway 05R at Taipei instead of 05L.

4

Clouded Vision

The Boeing 747 is the world's most prolific big jet. It was such a success from the beginning that Boeing announced in May 1985 that it was going to develop the 747-400 series with extended capacity and range. Since May 1990, the -400 has been the only 747 marketed, and the 1,000th 747 was rolled out on 10 September 1993 for delivery to Singapore Airlines. The advanced 747-400 long-range airliner was ideal for flying passengers in comfort across long distances, and on Tuesday 31 October 2000, a 747-400 (9V-SPK) belonging to Singapore Airlines departed Changi Airport, Singapore, for Los Angeles via Taipei.

Scheduled departure time for Singapore Airlines Flight 006 from Chang Kai Shek International Airport, Taipei, was 22.55hrs. 9V-SPK left Gate B-5 and taxied to Taxiway NP, which runs parallel to Runway 05L and 05R. The crew had been cleared for a Runway 05L departure because Runway 05R was closed owing to construction work. After reaching the end of Taxiway NP, SQ006 turned right into Taxiway N1 and immediately made a 180° turn to Runway 05R. After some six seconds holding, Flight SQ006 started its take-off roll at 23.15:45hrs. Weather conditions were very poor because Typhoon 'Xiang Sane' was passing through the area.

Just as the 747 was taking off, 3.5 seconds after V1, the aircraft commander Capt Chee Kong Foong saw a big concrete block straight ahead that had been put in place to close off part of the runway being repaired. It was too late to abort, and SQ006 hit the concrete block which brought it crashing back on to the runway. The jumbo jet burst into flames before sliding into more construction equipment, including two cranes, scattering wreckage across the runway. Aircraft wreckage was distributed along Runway 05R beginning at about 4,080ft from the runway threshold. The 747's 70m-long fuselage, now in two main sections, came to rest about 6,480ft from the runway threshold. Out of the 20 crewmembers and 179 passengers on board, four crew and 79 passengers perished.

Capt Chee Kong Foong and his two first officers — they were there to share the strain of the long haul across the Pacific — survived to give their version of events. In a region where face is everything, the managing director of Taiwan's Aviation Safety Council lost little time in releasing the last four minutes of taped conversation between Capt Foong, his co-pilot, and ATC:

23:15.18hrs	ATC (tower): 'Singapore 6, Runway 05 left [5L]. Wind 020 [degrees] at 28 [knots]. Gust to 50. Clear for take-off.'
23:15.26	Captain: 'Clear for take-off. Runway 05 left. Singapore 6.'
23:16.19	Captain: 'We can see the runway not so bad. OK, I am going to put it to high first.'
23:16.51	First Officer: '80 knots.'
23:16.52	Captain: 'OK, my control.'
23:17.08	First Officer: 'V1.' (142kt)
23:17.12	Captain: '(Expletive). Something there.'
23:17.13	Banging sound.
23:17.14	Captain: Unintelligible words, followed by a series of crashing sounds.
23:17.18	Recording stopped.

The tragedy happened because instead of taking off from Runway 5L as directed and acknowledged, the crew of the Singapore Airlines flight took off from the parallel runway 5R that was closed for repairs. The Taiwan Civil Aviation Authority had issued a Notice to Airmen (NOTAM) on 31 August 2000 indicating that part of Runway 05R between Taxiway N4 and N5 would be closed for construction work between 13 September and 22 November 2000. The aim of this work was to convert Runway 05R into Taxiway NC from 1 November 2000. The part of the runway where Foong took his wrong turn had been kept open for jets to taxi on.

How did such a dreadful mistake occur that ended with Singapore Airlines' first fatal disaster in 28 years of operation? The weather was an obvious culprit given that there had been a major storm that night. But both aircrew and local ATC personnel were well used to operating in typhoon-induced conditions, and other aircraft had taken off and landed around the time of the accident. The Taiwanese authorities said that the recorded visibility just after the crash was approximately 600m in heavy rainfall. A warning light in front of the construction site was on that night, but the pilots might not have been able to see it in the circumstances. Certainly, the weather must have played a factor, especially as the Taipei air traffic controllers were unable to see the 747 from the control tower. Standing on the crash site, it was obvious from the debris and mangled cranes that the crew of SQ006 had mistakenly lined up for take-off on the wrong runway.

There were certainly human error considerations involved. The knee-jerk reaction was to blame the pilots, and the 747's three-man flight crew were held by Taiwanese prosecutors, with the threat of criminal charges hanging over them for seven weeks, until pilots from more than 30 countries threatened to boycott Taipei airport if Capt Foong and his colleagues were not released.

The flight crew of SQ006 certainly missed or failed to appreciate some key warnings, including the NOTAM warning of the 05R closure and the two big signs indicating the 05R runway they mistakenly went down was closed. But if they made errors, they were not alone. The threshold of any runway closed for take-off and landings should always be marked with a big white cross to indicate to any pilot on the approach or on the ground that it was out of use, but there was

none on 05R. One runway marker light was broken, and another was not bright enough. The fact that part of 05R was officially open for taxying only muddied the waters. The Taipei airport authorities were as responsible for safe operations as the pilots using their facilities.

But the flight safety buck has to stop with the aircrew. The SQ006 accident was probably caused by a lethal mixture of confusion over what parts of 05R were open, inadequately delineated no-go areas shrouded in darkness and rain, and haste to get 159 paying passengers on their way to Los Angeles before the typhoon got any worse. There is always the diffidence factor on a big flightdeck. The captain oozes experience and the first officers, who may be much younger, see something untoward but are afraid to say anything for fear of looking stupid or getting their heads bitten off. That diffidence factor is even more prevalent on eastern airlines, where to imply that the great man has got it wrong is as bad as the great man admitting that he has screwed up in front of subordinates. It is hard to believe that everyone on the flightdeck of SQ006 was 100% sure that Runway 05R was 05L, but maybe no one felt confident enough to express any concern. The best aircraft commanders are those who encourage everyone on the flightdeck to sing out if they see anything going wrong, and the best airlines are those that don't penalise aircrew who admit to having made mistakes. Above all, saving face must never take precedence over flight safety, and both aircrew and ground staff should be doubly alert for misunderstandings and misinterpretations when the weather is foul.

Above:
A section of fuselage from SQ006 on the runway.
Associated Press

But if bad weather could present problems on flat airfields, it could be doubly lethal when it masked mountains with hard centres. Way back, when Bill Boeing was establishing his company in Seattle, Bill Piper over in Pennsylvania was dreaming that aeroplanes would become as common as automobiles. At a time when aircraft were painstakingly built virtually by hand, William T. Piper learned from innovators like Henry Ford that automobile assembly line techniques could increase productivity and decrease costs. From this came the affordable Piper Cub, selling initially for only $999 and becoming known as the aircraft that taught the world to fly.

In 1937 the Piper Aircraft Corporation moved to an old silk mill in Lock Haven, Pennsylvania, and after building tube-frame and fabric-covered aircraft for 17 years, Piper introduced its first all-metal aircraft, the Apache, in February 1954. It was also Piper's first twin-engined model and the first in a series to be named after an American Indian tribe as a salute to Piper's own Native American heritage. In 1967 the twin-engine line-up was enhanced with the introduction of the powerful, cabin-class PA-31 Navajo, designed to meet the growing demands for business travel. A series of descendants evolved from the PA-31, including a lengthened version called the PA-31-350 Chieftain, which could carry up to 10 passengers in some comfort.

A high spot for many tourists visiting Hawaii is a sightseeing flight around the 'Big Island'. A popular carrier is Big Island Air, whose publicity announces that 'We offer a complete tour of the Big Island. Destinations include the active Kilauea Volcano, lush green valleys, cascading waterfalls, black sand beaches and 266 miles of coastline. Tours are available from Keahole-Kona airport and start at $135. From Volcano Super Saver to Circle island tour, see the Big Island in our reliable, twin-engine, air-conditioned aircraft, complete with stereo music and pilot narration. Every seat is a window seat.'

On 25 September 1999 at about 16.22hrs Hawaiian standard time, Piper Chieftain (N411WL) owned by Big Island Air departed the Commuter Terminal at Keahole-Kona International Airport (KOA), Hawaii. It was the pilot's second sightseeing tour of the day, and Big Island Air Flight 58 was scheduled to follow a half-island tour route. Around 17.20hrs, the pilot sought permission from the Honolulu Automated Flight Service Station (AFSS) to transit through a restricted area of airspace that encompassed part of the centre of the saddle area between the Mauna Kea and Mauna Loa volcanoes. He was advised by AFSS that the restricted area was 'open', and the pilot was authorised to transition the valley-like area with its maximum elevation of about 6,800ft above mean sea level (amsl) for the next 30 minutes. No further radio transmissions were heard from the pilot. At 17.26hrs Flight 58 crashed at a point 10,100ft up on the northeast slope of the 13,330ft-high Mauna Loa volcano. The pilot and all nine passengers on board were killed.

Piper Chieftain N411WL had accumulated 4,523 flying hours since it was manufactured in 1983, and records showed that it had been maintained to approved standards. It was destroyed on impact with the lava-covered, up-sloping terrain and in the severe post-impact fire. All wreckage was found within an area approximately 150ft in diameter, including the flight control surfaces and structural components, so they had not dropped off beforehand. The Chieftain

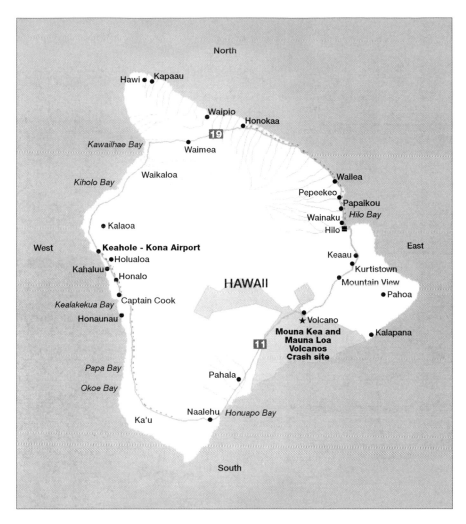

North

Hawi ● ●Kapaau

Waipio
●
Honokaa
●
19
Kawaihae Bay   Waimea
●

Kiholo Bay   Waikaloa
Wailea
●
Pepeekeo
●
Papaikou
●
Wainaku Hilo Bay
● Kalaoa
Hilo ▪

West   **Keahole - Kona Airport**   East
●
●Holualoa   Keaau ●
Kahaluu●
●Honalo   Kurtistown
●
HAWAII   Mountain View
●
● Pahoa
●Captain Cook
Kealakekua Bay
Honaunau●   ★ Volcano
Mouna Kea and   ● Kalapana
Mauna Loa
11   **Volcanos**
Crash site
Papa Bay
Pahala
Okoe Bay   ●

Naalehu Honuapo Bay
Ka'u   ●

South

was found inverted with its right wing, empennage, fuselage remains and both engine nacelles resting on the lava. The main landing gear was retracted, and no indication of in-flight fire was noted. No evidence of failures or in-flight malfunctions of the aircraft's systems was found, and there was no indication of mechanical malfunction or fire in either engine before impact. In sum, N411WL did not cause the crash.

Keahole-Kona International Airport was on the west side of the island. Big Island Air had three standard flight plans out of KOA filed with the Honolulu AFSS. Two went clockwise or anticlockwise round the circumference of the Big Island of Hawaii, while the third flew east-to-west to the opposite shore via the saddle area between the Mauna Kea and Mauna Loa volcanoes. The pilot's standard route for the accident flight would typically have been a circumferential route,

Above:
Map of Hawaii
— the Big Island

flying north from KOA to the Upolu Point at 1,500ft above ground level (agl). From there the pilot would head eastbound along the north shore toward the Hilo area at 1,000ft agl. From Hilo, the pilot would proceed south toward the Puu Oo vent and then to the large caldera crater of Kilauea. The pilot would then fly around South Point, the southern end of the island, and back to KOA along the western shoreline.

But frequently Big Island Air's pilots did not fly this designated circumferential route around the entire island. Rather than fly from the Hilo area to KOA via South Point, pilots sometimes flew back to KOA along the saddle route. Big Island Air did not have a flight plan covering this half-island tour route, but it was company policy that pilots could make this decision during flight because of meteorological conditions, volcanic activity or tourist needs. It was company policy for pilots to reverse course, as required, to avoid going IMC.

Federal regulations lay down that when operating VFR air tour flights over Hawaii, no pilot may come below an altitude of 1,500ft agl except during take-off, landings and other specific instances. Big Island Air had received FAA dispensation allowing its tour pilots to reduce the altitude flown at specific locations and transition segments to as low as 1,000ft agl when specific conditions were met. During such flights, aircraft had to remain at least 500ft below clouds, maintain 5km flight visibility, and remain within 0.5nm of the centreline of the approved flight route.

According to recorded radar data, at 17.21:04hrs the Chieftain was on the eastern side of the restricted area, about 8.5nm from the crash site. It was heading west back to KOA at an altitude, as transmitted by its Mode C-equipped transponder, of 6,600ft, and the underlying terrain elevation was about 4,600ft amsl. Both aircraft altitude and terrain elevation were increasing. Between then and the last recorded radar return at 17.25:29 when the Chieftain was within 0.5km of the accident site, Big Island 58's average ground track was 291° Magnetic, and its altitude increased from 6,600ft to 9,600ft. Several witnesses observed clouds below the 10,000ft-level around the time of the accident. At about 17.26, N411WL crashed on the northeast slope of the Mauna Loa volcano at 10,100ft amsl.

The Vice President of Big Island Air first met the deceased pilot in 1987 when they were both line captains for another air tour company in Hawaii. On the Vice President's personal knowledge about the pilot's flying background, his accident/incident-free flight record, his extensive flight experience in multi-engine aircraft, his PA-31-350 flight time, and his knowledge of Hawaii and its weather patterns, the pilot was hired by Big Island Air in May 1994. The pilot did not fly for Big Island Air between 1 September 1998 and 14 August 1999 because he went to study in Japan. On 30 August 1999, 25 days before the accident flight, the pilot re-certified after his 11-month leave of absence. During the FAA-administered flight check, the pilot was taken through the operating rules for VFR-only air tour flights over Hawaii. His flying skill level was assessed as good but, notwithstanding his total of 11,500 flight hours, the pilot was not authorised to perform instrument flight rules (IFR) air taxi flights. His FAA first-class medical certificate was issued on 14 August 1999, with the limitation that he wore corrective lenses. He did not use medication and was 'in good health'.

The pilot lived on the island of Oahu, from where he commuted to work by air. His landlord indicated that the night before the accident flight, the pilot went to bed about 21.00hrs. He was scheduled to make two air taxi flights on 25 September, the first departing about 07.00hrs and the accident flight at about 16.22. Given the early morning schedule, the landlord believed it likely that the pilot would have risen about 04.00hrs, driven to Honolulu International Airport and flown on a commercial flight to KOA to begin work.

A Big Island Air employee who saw the pilot on his arrival at KOA on the day of the accident stated that the pilot appeared to be alert and well rested. Big Island Air pilots were responsible for obtaining weather information before departure, but no one provided the pilot with weather information for the accident flight, and there was no record of the pilot requesting a weather briefing for either of his flights on 25 September as required by the FAA.

There was no continuously operating aviation weather reporting facility in the saddle area of the Big Island. On the eastern (windward) side of the island, clouds often form over the up-sloping terrain. The skies on the western side of the island are typically clear or have scattered clouds, which partly explains why KOA was built there.

Three US Geological Survey employees observed weather conditions about 2.5km northeast of the accident site on the eastern slope of the Mauna Loa volcano on 25 September. Two of them indicated that at about 14.45, visibility was 30-200m and the sky was 'mostly closed', but it began clearing about 16.30. Later, around the time of the accident, a column of grey smoke was seen to the southwest. The witnesses indicated that the visibility on the far side of the smoke was 'murky'. Although the smoke column was in clear skies, a witness indicated that the landscape around the area of the smoke column was not well defined because of the clouds. The winds were reported as being very slack at that time.

A helicopter pilot on the prescribed route at about 16.30hrs reported that to the north the sky was overcast, with ceilings below 500ft above ground level. To the south of the route, an overcast cloud layer between 5,000ft and 7,000ft amsl extended from the east flank of the Mauna Loa volcano towards the ocean.

Mauna Loa Observatory is perched at 11,140ft amsl and about 6nm from the accident location. On 25 September at about 17.00hrs, the temperature was 8.8°C, wind was calm, and the dew point was -13.9°C. Several cameras were fixed on a 63ft tower at the observatory, and imagery facing north, east and southwest taken at 10min intervals around the time of the accident showed fog conditions around the observatory, with part of the ridgeline of the volcano visible in several photographs.

From all the evidence that the sky was overcast in the vicinity of the accident site, NTSB investigators concluded that on the afternoon of 25 September the pilot flew into IMC in the vicinity of the accident site. Investigators determined that the probable cause of this accident was the pilot's decision to continue visual flight into instrument meteorological conditions in an area of cloud-covered mountainous terrain. Contributing to the accident were the pilot's failure to navigate properly and his disregard for standard operating procedures, including flying into IMC while on a visual flight rules flight plan and failure to obtain a current pre-flight weather briefing.

Big Island Air's FAA-approved Operations Specifications and corresponding training programme clearly stated that all tour flights were to be conducted under VFR. No flying under instrument flight rules was authorised at any time. The properly certificated and qualified pilot had over 11,500 hours of flight time, most of which was accumulated in the Hawaiian islands, so he must have been well aware of the VFR flight visibility and cloud clearance limitations.

The pilot was responsible for obtaining a pre-flight weather briefing from the FAA's Honolulu Automated Flight Service Station as required by the FAA. The FAA has no record of the pilot requesting a weather briefing for the accident flight or the flight he conducted earlier that day. The Safety Board concluded that the pilot's failure to obtain a pre-flight weather briefing was a deviation from standard operating procedures.

The accident Chieftain was equipped with very high frequency omnidirectional range, distance measuring equipment and global positioning satellite receivers, which could have been used to indicate the air taxi aircraft's position and ground clearance. Although the pilot was not required to use these navigational aids, when he departed visual meteorological conditions and flew into IMC he should have used the navigational aids to monitor accurately his ground track and altitude. During the last few minutes of flight, when the aircraft's ground clearance was rapidly decreasing, the pilot did not reverse course or take emergency action. Radar data indicated that at this part of the flight, N411WL's track varied little from its predominantly west-northwesterly direction. If the pilot had been using his navigational aids correctly, he would have realised that he was nearing high terrain and would probably have changed his course. Pilots can often get caught out by rapidly deteriorating weather, but they should always be aware of their position so that they can escape from potential trouble safely and expeditiously.

There were other human factors in an accident that was more than just about weather. Two days after the accident, the Safety Board received correspondence from a passenger who had flown with the accident pilot during a Big Island Air sightseeing tour on 4 September 1999. The passenger's description of the flight route was similar to that of the accident flight. The passenger indicated that during his tour, the pilot had flown in dense clouds that prevented him from being able to see both ahead of and below the aircraft. Investigators subsequently viewed photographs and a video taken by the passenger. The latter showed the Chieftain flying in clouds on several occasions and in different locations throughout the flight.

The fatal run into the volcano occurred less then 30 minutes before sundown. As he had been up since 4am and been working a long duty day, fatigue could have been a factor in this accident. Furthermore, the pilot's immediate superior, Big Island Air's chief pilot, had never flown with the pilot. Knowing your aircrews' competencies and their fitness to fly should not be left to an annual check flight. And this is not just a management issue. Surely someone else in Big Island Air had got to know that the pilot had a tendency to sail close to the wind.

Years ago the Squadron I had just left lost a big, four-engined bomber and six aircrew lives. When you throttled two engines back to practise an asymmetric approach, you had to overshoot at decision height or land. To try and overshoot below decision height by opening up all four engines was forbidden because the two at idle would be much slower to respond than the other two. Most of the time you could get away with it but on this occasion two engines on one side got to be pushing out 40,000lb of thrust on one side with next to nothing to offset this on the other. As no captain could have held this manually, the great V-bomber turned on its back and crashed in the middle of the airfield. Afterwards, co-pilots said that the captain had got into the habit of practising asymmetric overshoots from below decision height. If we had all been willing to speak up before the accident, six men might still be alive. All aircrew have a responsibility to ensure that the rules are not bent.

*   *   *   *   *

The Douglas Aircraft Company delivered 976 twin-jet DC-9 airliners between December 1965 and October 1982. A higher capacity variant then appeared, known as the McDonnell Douglas MD-80, followed by numerous versions including the MD-82 designed for hot and high performance and increased payload/range.

On 1 June 1999 at about 22.40hrs, American Airlines Flight 1420 departed Dallas/Fort Worth International Airport, Texas, for Little Rock, Arkansas. On board MD-82 N215AA were two pilots, four flight attendants and 139 passengers, and they touched down at Little Rock National Airport at 23.50:20hrs Central Daylight Time. But instead of stopping normally, N215AA overran the end of Runway 04R. Exactly 411ft later it struck several tubes extending outward from the left-hand edge of the ILS localiser array. N215AA then passed through a chain-link security fence and over a rock embankment to the flood plain approximately 15ft below, before colliding with the structure supporting the Runway 22L approach lighting system. The captain and 10 passengers were killed. The first officer, the flight attendants and 105 passengers received serious or minor injuries. Only 24 passengers were uninjured. The MD-82 was destroyed by impact forces and in the post-crash fire.

Flight 1420 was the third and final leg for the pilots. Their flight sequence began at Chicago O'Hare where the captain checked in about 10.38hrs, some 30 minutes after the first officer. Flight 1226, from Chicago to Salt Lake City, Utah, departed about 11.43hrs and arrived just over three hours later. The next leg, as Flight 2080 from Salt Lake City to Dallas/Fort Worth (DFW), departed about 16.47 and arrived around 20.10hrs, 39 minutes behind schedule after being held off by adverse weather in the airport area. The captain flew Flight 1226, and the first officer was the flying pilot for Flight 2080.

Flight 1420 to Little Rock was scheduled to depart at about 20.28 and arrive at about 21.41. However, before their arrival at DFW, the flight crew were told that the departure of Flight 1420 had been delayed to 21.00hrs. After disembarking from Flight 2080, the flight crew went to the departure gate for Flight 1420 where they received flight planning paperwork including several weather advisories for widely scattered and severe thunderstorms along the planned route.

A severe thunderstorm is defined as having winds of 50kt (58mph) or greater, ¾in or larger hail, and creating torrential rain and frequent lightning. The MD-82 originally slated for the flight was delayed in its arrival at DFW, and after

21.00hrs the first officer notified gate agents that Flight 1420 would need to depart by 23.16 otherwise the flight crew would be outside American Airlines' maximum pilot duty time, which was set at 14hr from the scheduled time of check-in for the first flight leg to the time that the last flight leg was scheduled to land. N215AA was then substituted, but the upshot was that when the MD-82 finally got airborne at 22.40hrs, Flight 1420 was 2hrs 12min behind the scheduled departure time.

The captain was the flying pilot and at about 22.54 the flight dispatcher sent a message suggesting an expedited arrival at Little Rock to beat predicted thunderstorms. The first officer recalled that 'there was no discussion of delaying or diverting the landing' because of the weather, but the pre-departure flight paperwork had specified Nashville International and DFW as options in case Flight 1420 had to divert.

At about 23.04 Fort Worth Center broadcast a Weather Advisory for an area of severe thunderstorms that included Little Rock airport. At 23.25:47 the captain stated, 'We got to get over there quick.' About 5 seconds later, the first officer said, 'I don't like that . . . that's lightning,' to which the captain replied, 'Sure is'. The flight crew had the city of Little Rock and its airport area in sight at 23.26:59.

The flight crew contacted Little Rock ATC at 23.34:05hrs. The Tower controller advised that a thunderstorm located northwest of the airport was moving through the area and that the wind was 280° at 28kt, gusting to 44kt. The first officer told ATC that they could see the lightning, and the pilots were told to expect an ILS approach to Runway 22L. During descent into the terminal area, the stormy weather appeared to be about 24km away from the airport and the pilots thought that they had 'some time' to make the approach.

At 23.39:00 the controller cleared the flight to descend to 3,000ft. At 23.39:12 the first officer stated, 'Okay, we can . . . see the airport from here. We can barely make it out but we should be able to make [Runway] Two Two . . . that storm is moving this way like your radar says it is but a little bit farther off than you thought.' ATC offered Flight 1420 a visual approach to the runway, but the first officer indicated, 'At this point, we really can't make it out. We're gonna have to stay with you as long as possible.'

At 23.39:45 the controller notified Flight 1420 of a wind shear alert, reporting that the centrefield wind was 340° at 10kt, the north boundary wind 330° at 25kt, and the northwest boundary wind 010° at 15kt. The pilots then requested Runway 04R so that they would have a headwind, rather than a tail wind, during landing. ATC told the flight crew to fly a heading of 250° for vectors to finals for an ILS approach to 04R. On reaching the assigned heading, N215AA was turned away from the airport and clear of the reported thunderstorm.

Between 23.42:19 and 23.42:24 the captain asked the first officer, 'Do you have the airport? Is that it right there? I don't see a runway.' At 23.42:27, the controller told the flight crew that the second part of the thunderstorm was apparently moving through the area and that the winds were 340° at 16kt, gusting to 34kt. At 23.42:40 the first officer asked the captain whether he wanted to accept 'a short approach' and 'keep it in tight'. The captain answered, 'Yeah, if you see the runway, 'cause I don't quite see it.' The first officer responded, 'Yeah, it's right here, see it?' The captain replied, 'You just point me in the right direction and I'll

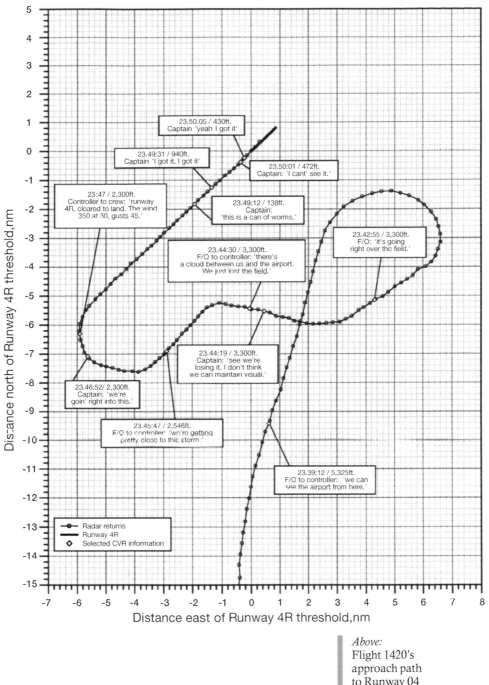

Chart axes:
- Y-axis: Dis-ance north of Runway 4R threshold,nm (from -15 to 5)
- X-axis: Distance east of Runway 4R threshold,nm (from -7 to 8)

Annotations on chart:

23.50.05 / 430ft.
Captain 'yeah I got it'

23.49:31 / 940ft.
Captain 'I got it, I got it'

23.50:01 / 472ft.
Captain: 'I cant' see it.'

23:47 / 2,300ft.
Controller to crew: 'runway 4R, cleared to land. The wind 350 at 30, gusts 45.'

23.49:12 / 138ft.
Captain: 'this is a can of worms.'

23.42:55 / 3,300ft.
F/O: 'it's going right over the field.'

23.44:30 / 3,300ft.
F/O to controller: 'there's a cloud between us and the airport. We just lost the field.'

23.44:19 / 3,300ft.
Captain: 'see we're losing it. I don't think we can maintain visual.'

23.46:52/ 2,300ft.
Captain: 'we're goin' right into this.'

23.45:47 / 2,546ft.
F/O to controller: 'we're getting pretty close to this storm.'

23.39:12 / 5,325ft.
F/O to controller: 'we can see the airport from here.'

Legend:
- Radar returns
- Runway 4R
- ◇ Selected CVR information

Above:
Flight 1420's approach path to Runway 04 at Little Rock, together with key CVR comments.

start slowing down here.' At 23.42:55 the first officer said, 'It's going right over the . . . field.' At 23.42:59, the first officer told the controller, 'Well, we got the airport. We're going between clouds. I think it's right off my, uh, three o'clock low, about four miles.' The controller then offered a visual approach to Runway 04R, and the first officer accepted.

At 23.43:35 the first officer told the captain, 'You're comin' in. There's the airport.' Three seconds later, the captain stated, 'Uh, I lost it,' to which the first officer replied, 'See, it's right there.' The captain then stated, 'I still don't see it . . . just vector me. I don't know.'

At 23.43:59 the controller cleared the MD-82 to land and indicated that the winds were 330° at 21kt. At 23.44:19 the captain stated, 'See, we're losing it. I don't think we can maintain visual.' At 23.44:30 the first officer told ATC that visual contact with the airport had been lost because of a cloud between the airliner and the airport. The controller then cleared Flight 1420 to fly a heading of 220° for radar vectors for the ILS approach to Runway 04R and directed it to descend to and maintain 2,300ft.

At 23.45:47 the first officer told the controller, 'We're getting pretty close to this storm, we'll keep it tight if we have to.' The controller replied that, 'When you join the final, you're going to be right at just a little bit outside the marker if that's gonna be okay for ya.' The captain stated, 'That's great,' and the first officer told the controller, 'That's great with us.' At 23.46:39 the controller advised the flight crew that the airliner was 5km from the ILS outer marker, located just under 9km from the airport. The first officer subsequently stated that, at this point, there was urgency to land because the weather was 'up against' the airport.

At 23.46:52, the captain stated, 'Aw, we're goin' right into this.' At the same time, the controller reported that there was heavy rain at the airport, visibility was less than 1.5km, and the runway visual range (RVR) — the measurement of visibility near the runway's surface from the approach end — for 04R was 3,000ft. The first officer acknowledged this transmission. At 23.47:08 the controller again cleared Flight 1420 to land and indicated that the wind was 350° at 30kt, gusting to 45kt. At 23.47:22 the captain stated, 'Three thousand RVR. We can't land on that.' Four seconds later the first officer said that the approach chart showed that the lowest authorised RVR for that runway was 2,400ft, whereupon the captain then said, 'Okay, fine.'

At 23.47:44, the captain asked for 'landing gear down', and the CVR recorded the sound of the landing gear being operated. About 5 seconds later the captain stated, 'And lights, please.' At 23.47:53 the controller issued a second wind shear alert for the airport, reporting that the centrefield wind was 350° at 32kt gusting to 45kt, the north boundary wind was 310° at 29kt, and the northeast boundary wind was 320° at 32kt. The flight crew did not acknowledge this transmission. At 23.48:10 the captain stated, 'Add twenty [knots],' to which the first officer replied, 'Right.'

At 23.48:12 ATC reported that the RVR was now down to 1,600ft. Six seconds later the captain said that he was established on final approach; after another six seconds, the first officer informed the controller that the flight was established on the inbound portion of the ILS. The controller repeated the clearance to land, stating that the wind was 340° at 31kt, the north boundary wind was 300° at 26kt, and the northeast boundary wind was 320° at 25kt, and he repeated the RVR. The first officer acknowledged this information but the controller did not receive any further transmissions from Flight 1420. At 23.49:02, the first officer asked the captain, 'Want forty flaps?' The captain thought he had already called for the landing flaps, after which the first officer stated, 'Forty now'. At 23.49:10 ATC reported the crosswind as 330° at 28kt. Two seconds later, the captain stated, 'This is a can of worms.' The first officer stated, 'There's the runway off to your right, got it?' The captain replied, 'No,' to which the first officer stated, 'I got the runway in sight. You're right on course. Stay where you're at.' The captain then stated, 'I got it. I got it.' At 23.49:32, the controller reported the wind to be 330° at 25kt. At 23.49:37 an unidentified voice in the cockpit stated, 'Wipers,' and the CVR recorded a sound consistent with windshield wiper motion. The wind at 23.49:53 was reported as 320° at 23kt.

At 23.49:57 an unidentified voice in the cockpit stated, 'Aw . . . we're off course' and then, a second later, an unintelligible comment was made by an unidentified voice. Afterwards, the first officer stated that he thought the approach was stabilised until about 400ft above the field, at which point the airliner drifted to the right. The first officer said 'go around' about that time but not in a very strong voice. He looked at the captain to see if he had heard him but that the captain was intent on flying and was doing 'a good job'.

At 23.50:00 the first officer said, 'We're way off' and ILS localiser deviation at the time was about one dot to the right. After a second the captain stated, 'I can't see it.'

Three seconds later the first officer asked, 'Got it?' to which the captain replied, 'Yeah, I got it.' Shortly afterwards the CVR recorded the sound of the ground proximity warning system alerting the pilots that their rate of descent was excessive.

The MD-82 was descending through 70ft agl at the time of the first sink rate warning, and through about 50ft at the second.

N215AA touched down on the Little Rock runway at about 23.50:20hrs. At around 23.50:22 the first officer stated 'We're down' and about two seconds later he stated, 'We're sliding.' Over a seven-second period after touchdown, both thrust reversers were deployed. Right and left brake pedals began to move at 23.50:25, and both pedals reached full travel at 23.50:31. About the time that the brakes were applied, the thrust reversers were deployed again. At 23.50:32 the CVR recorded an unidentified voice in the cockpit calling 'on the brakes'. The left brake pedal was relaxed at 23.50:34 before returning to its full position 2 seconds later. About the time that the left brake pedal was relaxed, the reversers were returned to the unlocked status.

At 23.50:40 the left thrust reverser was moved back to the deployed position, but the right reverser moved briefly to the deployed position and then moved to the stowed position. The left thrust reverser remained deployed, and the right thrust reverser remained stowed, for the remainder of the flight. About 1 second later, the CVR recorded expletives by an unidentified voice in the cockpit, followed by the sounds of initial impact at 23.50:44 and several additional impacts three seconds later. The CVR stopped recording at 23.50:48, by which time the airliner had come to rest about 800ft from the departure end of Runway 04R. Damage to the airfield structure was estimated at $325,000.

N215AA, which had been delivered new to American Airlines on 1 August 1983, had accumulated 49,136 flight hours by the time it hit Little Rock. The fuselage separated into three main sections, the left wing was fractured and completely severed near its root and wingtip, and the nose gear and right main landing gear were sheared from their attachments. The collision with the approach lighting system crushed the aircraft nose rearward, while destroying the left side of the fuselage back to the cockpit rear bulkhead and from the beginning of the first-class section aft to the second row of the coach section. The collision with the approach lighting system also created a hole in the left side of the cabin that extended from the overhead stowage bins to the cabin floor in the first-class and coach sections.

The first officer could not evacuate N215AA on his own because his left femur was fractured. He had to be removed by rescue workers, who had to cut through metal and then step onto the centre pedestal to extricate him. Rescue workers also had to remove some surviving passengers from the first-class section. The flight attendants seated on the forward jumpseat were seriously injured in the crash and could not assist with passenger evacuations.

Of the passengers who were forward of the fuselage separation, 18 escaped through a large hole on the left side of the first-class section or through a separation in the fuselage. The forward entrance and forward galley doors could not be used because of structural deformation of the fuselage. Six passengers seated on the left side of the first-class section were ejected in their seats through the large hole, while another seven were ejected in their seats into the area between the fuselage sections. Fortunately, nine of these 'ejectees' survived. The flight attendant seated on the inboard forward jumpseat was carried out by a passenger through the large hole in first class.

All four overwing emergency exits were opened by passengers from inside the cabin. The passenger seated next to the left aft overwing exit was able to open the hatch, reporting that the door 'seemed to pop out easily and quickly'. The flight attendant seated in the aft cabin jumpseat opened the aft bulkhead door (leading to the tail cone exit) with the assistance of passengers. The flight attendant and several passengers entered the tail cone area, but the tail cone did not fall away from the airliner after the flight attendant and at least one passenger pulled the release handle. The flight attendant and passengers then kicked and jumped on the tail cone and created a gap between the fuselage and the tail cone that 12 people used to escape.

From the airport firefighters' point of view, crashes in bad weather and darkness are always chaotic affairs, and N215AA's arrival at Little Rock was no exception. ATC called the Aircraft Rescue and Fire Fighting (ARFF) units on the crash phone at about 23.52hrs. The local controller flagged up the possibility of an accident at the end of 04R but did not specify which end. The ARFF units went to the approach end of 04R, but the MD-82 had gone off the departure end. As a result, the ARFF units had to travel back to the taxiway and then go to the other end of the runway. They located the remains of N215AA about 11 minutes after the initial call from the local controller.

However, they did not arrive on the scene for another five minutes because they had to travel in the opposite direction to an access road, turn on to a perimeter

road back in the direction of the accident site, stop to unlock a perimeter security gate manually, and then continue on the perimeter road to the accident site. If ARFF units had known the approximate location of the stricken airliner when they left the fire station, much time and trouble would have been saved.

The accident was not survivable for the captain and the passengers seated on the forward left side of the airliner where it collided with the approach lighting structure, plus those immediately exposed to lethal impact forces or fire where the fuselage separated. The accident, however, was potentially survivable for the passenger fatalities in seats 27E and 28D. Investigators considered whether a shorter ARFF response time could have prevented these fatalities but, even with the shortest possible response time, the passenger in seat 28D would have already received the second- and third-degree burns to over half of her body and the severe inhalation injury from which she later died.

So why did the MD-82 overrun Runway 04R to such catastrophic effect? The captain and the first officer were properly certificated and qualified, and neither N215AA nor the Little Rock runway was at fault. The weather was bad but thunderstorms and wind shear did not of themselves cause this accident. They just triggered a number of human factors.

Taking the accident in stages, the pilots had plenty of warning of thunderstorms during the approach. They had information from their airborne weather radar and they would have been able to see some of the 46 cloud-to-ground lightning strikes detected by the National Lightning Detection Network within five miles of the airport in the five minutes before the accident. The CVR showed that the pilots discussed the weather and the need to expedite the approach.

Above:
An American
Airlines MD-82.
*The Aviation
Picture Library*

American Airlines did not prohibit its flight crews from continuing an approach so long as their intended route was clear of the thunderstorms. In his subsequent testimony, the first officer stated that, during the descent, the weather was to the left and moving off to the right and that the airport looked clear. Thus, during the descent into the terminal area, the pilots could have reasonably believed that they could reach the airport before the thunderstorm.

The crew of Flight 1420 had been told by ATC to expect an ILS approach to Runway 22L. The local controller broadcast the first of two wind shear alerts when the MD-82 was about 12km from the airport. As the winds had shifted to the northwest, the flight crew requested a change of runway to 04R to land into wind. At this stage, the leading edge of a line of thunderstorms — one of the most hazardous areas because of the up-draught/down-draught interaction — was over the airport, with the heaviest activity located northwest through north-east 8km from the runway.

ATC vectored Flight 1420 right to Runway 04R. However, this turn, and sub-sequent manoeuvring south of the airport, meant that the pilots were temporarily unable to use their forward-looking airborne weather radar to monitor the storm's intensity and location relative to the airport. They would regain the use of the weather radar only when the MD-82 turned back to intercept the final approach for 04R. For about 7 minutes the crew did not have precise radar infor-mation about the location of the storm relative to the airport.

American Airlines' flight manual stated that an aircraft commander was to ensure that a briefing was conducted before every approach, but it did not mention an abbreviated briefing for an instrument approach to a changed run-way. There was no discussion between the pilots about the missed approach procedure in relation to the location of the storm. They continued to receive but did not discuss wind reports from ATC. The pilots' actions and conversa-tions during this phase of the approach consisted largely of steps to expedite the approach to the airport.

The ATC controller offered a visual approach to 04R, which the crew accepted. During the attempted visual approach, the captain could not see the airport and relied on the first officer to guide him. It is arguable that, at this stage, the captain should have handed over control to the first pilot, who had flown the leg into Dallas Fort Worth, and who was best placed visually to get the airliner down onto the ground quickly.

At 23.47:08 the controller cleared Flight 1420 to land and stated that the crosswind was 350° at 30kt, gusting to 45kt. The pilots did not discuss the wind information, the heavy rain that was already falling at the airport or the depic-tion of the weather on the airborne radar. Their main concern was the location of the thunderstorm in relation to the airport and the aircraft. However, because the first officer was able to maintain visual contact with the runway, both pilots might still have believed that Flight 1420 could arrive at the airport before the thunderstorm.

Reports of heavy rain, deteriorating visibility and increasing crosswinds would have decided some flight crews to hold off until the storm passed or to divert to an alternate airport. These would have been wiser courses because American's maximum crosswind component for a MD-82 landing on a runway with an RVR less than 4,000ft was 15kt. The latest ATC winds gave a crosswind component of 23kt for the steady-state wind and 34kt in gusts. Thus, a landing on 04R could no longer be conducted in accordance with company procedures.

The captain called for landing gear down at 23.47:44. According to American's procedures, this callout indicated that the second half of the 'Before Landing' checklist — landing gear, spoiler lever, autobrakes, flaps and slats, and annunciator lights — was to be accomplished using a mechanical checklist in the cockpit. The CVR recorded the sound of the landing gear being operated two seconds after the callout, followed by the captain's statement 'And lights, please' three seconds later. At this point, none of the remaining checklist items had been performed.

At 23.47:53 ATC passed on the second of two wind shear alerts whereupon the captain wisely asked for an extra 20kt on the approach speed. There was no further discussion of the winds, which still exceeded American's maximum crosswind limitations for landing. The pilots should have discontinued the approach to 04R because the maximum crosswind component for landing had been exceeded.

At 23.48:12 the controller told the flight crew that the RVR had dropped to 1,600ft, which reduced American's maximum landing crosswind component to 10kt. However, there was still no discussion on the flightdeck about the exceeded maximum crosswind component. The first officer stated afterwards that he was still able to see the airport and that the RVR information provided by the controller 'did not concur' with what he and the captain were seeing.

When they started that final approach to Runway 04R, the pilots entered a high workload phase of flight which, under normal conditions, would include tasks such as controlling and manoeuvring the MD-82, configuring it to land, actioning final landing checks and evaluating the airliner's performance relative to the landing criteria. With Flight 1420, the pilots' decision to accept a short approach increased their already high workload by compressing the amount of time available to get through the required tasks. As the first officer recalled, 'I remember around the time of making that base-to-final turn, how fast and compressed everything seemed to happen.'

When the first officer called 'established inbound,' ATC cleared the MD-82 to land and provided updated winds. There were many cues, such as the heavy rain, the rapidly decreasing RVR and the shifting and gusting winds, to show that the weather at and around the airport had deteriorated substantially. However, there was no discussion between the pilots about whether to continue the approach. The deteriorating weather would have prompted many flight crews to abort the approach.

Flight 1420 touched down doing 160kt some 2,000ft down the 7,200-ft runway, slightly to the right of the centreline and sliding to the right. The MD-82 was subjected to a 5kt tail wind component on touchdown and a 20-25kt left-to-right crosswind component during the landing. N215AA ran off the end of 04R, sliding to the left at 97kt, and no damage was sustained until it collided with the 22L approach lighting support structure at around 83kt. There was no evidence of aquaplaning. Investigators concluded that the lack of spoiler deployment was the single most important factor in the flight crew's inability to stop the MD-82 within the available runway length. And the serviceable spoilers did not automatically deploy because neither pilot armed the spoiler handle before landing.

It is always easy to blame aircrew when things go badly wrong and passengers die. The 48-year-old captain of Flight 1420 was hired in 1979 after he left the USAF. He qualified as an MD-80 captain in July 1991 and by 1999 he was not only a lieutenant-colonel in the US Air Force Reserves but also chief pilot and check airman at Chicago O'Hare. The captain attained the chief pilot position because of his flying background, company achievements and leadership skills. When he died, the captain had amassed over 10,000 flight hours, more than half of which were as an MD-80 captain.

The pairing with the captain of the 35-year-old first officer, who was five months into his probationary year, was not a factor in the accident. Although a probationary first officer can be overawed by a chief pilot, CVR evidence indicated that this first officer was assertive during most of the flight. All the evidence pointed to these two pilots being dedicated and professional operators, and their performance into Little Rock was widely at variance with both pilots' normal skills, abilities and cockpit style.

But the conditions at Little Rock were not normal. Although the captain was known as a conservative aviator who used common sense and oozed wisdom and experience, and the first officer was an above-average new hire who was very competent and knowledgeable with good cockpit discipline, the pair fell into the trap of focusing to a fatal degree on the deteriorating weather and the consequent need to expedite the landing. Now that aerospace technology has become so reliable, the greatest pressures on aircrew are those imposed by themselves in response to commercial and human factors.

Decision-making can be degraded when stressed individuals selectively focus on just one subset of environmental cues. Consequently, any situation assessment may be incomplete, and the resulting decision, even when made by an expert, may be degraded. Stress can also impede an individual's ability to evaluate an alternative course of action, resulting in adherence to an original plan long after it is right or sensible.

The pilots' preoccupation with expediting the landing at Little Rock diverted their attention away from other activities during the final minutes of the flight and, as a result, affected their ability properly to assess the situation and make effective decisions. It did not help that it had been a long day. The CVR contained no statements to indicate that either pilot was tired, and there was no evidence that they had experienced cumulative sleep loss in the days before the accident. But at the time of the accident, just before midnight, the captain and the first officer had been continuously awake for at least 16 hours. Research indicates that the normal waking day is between 14 and 16 hours and lapses in vigilance increase in frequency and become longer if the normal waking day is extended. Fatigue and commercial pressure to get the already-delayed passengers down on the ground may have been the final reasons why the flight crew failed to discontinue the approach when severe thunderstorms moved into the airport area, and failed to ensure that the spoilers had extended after touchdown. The great lesson from the loss of Flight 1420 at Little Rock was that every airline's organisation, crew training system and culture must be sensitive enough to enable human aircrew to operate safely in the face of whatever the elements throw against them.

5

Near Miss

The Boeing 757-200 is a medium-range airliner powered by two large turbofan engines. Designed by Boeing to replace its 727, the 757-200 can carry 304 passengers in two classes up to 4,390 miles. The 757-204 version, ordered by Britannia Airways, makes an ideal workhorse for carrying passengers quickly, safely and cost-effectively on short- and medium-haul trips around Europe. Boeing 757-204 G-BYAN was delivered to Britannia Airways on 26 January 1994.

On 22 November 2000 G-BYAN was scheduled for a turnaround flight from Birmingham Airport to the tourist resort of Paphos on the southwest coast of Cyprus. The aircraft departed Birmingham's Runway 15 in a southerly direction and entered cloud in the climb between 3,000ft and 4,000ft. Not long after G-BYAN got airborne, the departure controller cancelled the Standard Instrument Departure procedure and placed the aircraft under radar control. At FL60 control was handed from Birmingham Departures to Midland Terminal Control (MTC).

Immediately upon contact with MTC the 757 was cleared to climb to FL90 and shortly thereafter given a radar heading of 140°. About one minute after initial contact with MTC the 757 was re-cleared to FL100. The controller acknowledged the B757 crew report on reaching FL100 and advised, 'Military traffic in your eleven o'clock position crossing left to right, one thousand feet above.' The 757 crew acknowledged the advice and although their aircraft remained in cloud, they immediately began a visual search for the traffic, which their Traffic Alert and Collision Avoidance System (TCAS) was indicating 1,000ft above. The cloud was too thick for visual contact with the military traffic, but the crew carried on looking out as the TCAS contact passed clear down their right side.

Shortly after the traffic passed clear and whilst still in cloud, the commander and the first officer suddenly became aware of an aircraft in their left 'half-past ten' position at very close range and at about the same level. The aircraft, which the airline pilots were immediately able to identify as a twin-fin fighter and later as an F-15, passed rapidly across the 757's nose and disappeared down their right side. The 757 crew heard the noise of the F-15's Pratt and Whitney engines and they felt its wake turbulence at 10.20hrs. There was no time for the 757 pilots to take avoiding action. Subsequent analysis of radar data indicated that at the closest point of approach the two aircraft were separated by less than the minimum range detectable by the radar, which for the technically inclined is 0.0625 of a nautical mile or about 380ft. None of the six flight attendants or 234 passengers saw the F-15, but the cabin crew felt the disturbance as the 155ft-long airliner flew through the F-15's wake. There were no injuries to anyone on board G-BYAN, so the flightdeck crew filed an air proximity (Airprox) report with ATC and continued to Cyprus.

While 234 tourists were preparing to leave Birmingham for the sunny Mediterranean, across the country in rural Suffolk life was getting busy at RAF Lakenheath. This air base. is home to the USAF's 48th Fighter Wing — the Statue of Liberty Wing — with three squadrons of F-15 Eagles. The twin-engine, twin-boom F-15 began life as a single-seat air defence interceptor, but this potent machine was subsequently given a second crew station and developed into a dual role attack/air superiority fighter. The F-15E Strike Eagle is probably the finest fighter bomber in the world and on the morning of 22 November 2000 two F-15Es started up on the Lakenheath flight-line tasked with providing flight training for the front seat occupant of the No 2 aircraft.

The pilot under training was in current flying practice on the single-seat F-15C, but as he needed to be brought up to speed in the ground attack role and there are several significant differences between the Eagle and the Strike Eagle, an F-15E instructor pilot rather than a weapons system officer was in the rear seat. The intention was to carry out tactical low-flying training over Wales followed by bomb delivery practice on one of the Wash air weapons ranges, before returning to base for circuit training.

A solid chunk of controlled airspace runs northwest-southeast up the centre of England. This is the spine of the UK air traffic system, flowing up to Scotland and thence across the Atlantic, or down to southern England and the Channel or Thames Estuary air lanes across to continental Europe. The F-15s needed to cut across what amounted to an aerial version of the M1 motorway measuring 50nm from edge to edge, and they could not just sneak across one the busiest and most rigidly controlled pieces of airspace in Europe in the hope that nobody noticed. One corridor had been set up at Lichfield to ease the path of military front-line traffic out of Lincolnshire, while another further south was geared to the East Anglian military airfields. The latter corridor was established at right angles to the controlled airspace centred on the Daventry VOR/DME orientated on the 066°/246° radials. The Strike Eagles planned to fly to Wales at medium altitude through the Daventry Radar Corridor, descending to low level once clear of controlled airspace to the west.

The two F-15s lifted off from Lakenheath approximately 20 seconds apart and took up a 'trail' formation with the No 2 aircraft about 3km behind the leader. In accordance with standard procedures for this type of two-ship formation, only the lead F-15 transmitted a Secondary Surveillance Radar (SSR) Squawk. The lead aircraft climbed through cloud, with his No 2 maintaining position by use of radar; they both levelled at FL100 in VMC. Part of the briefed flight profile included an aircraft systems check for both aircraft in VMC, which involved a change of lead aircraft. The No 2 completed his checks and began to close on his No 1 to take the lead position when the formation entered IMC. The No 2 aborted the change of lead and dropped back to about

a 2km in trail. In an attempt to regain VMC, the leader requested ATC clearance to climb to FL110.

Soon after the F-15s got airborne, Lakenheath Departure Control handed ATC control of the formation to London Military Radar. London Mil cleared the F-15s to cross the Daventry Radar Corridor at FL100, and the lead F-15's request to climb to FL110 was made to London Mil shortly after the aircraft entered the Daventry Radar Corridor. Initially, the controller told the flight to maintain FL100 while she contacted the MTC controller by landline to co-ordinate a climb. The MTC controller agreed the higher level and London then cleared the F-15s to climb to FL110.

The leader immediately began a climb to FL110, but the No 2 crew did not hear the ATC clearance and maintained FL100. The two pilots in the No 2 aircraft later noticed that their radar showed the leader to be above their level, which generated a bit of discussion. At about this time, the front seat pilot became vaguely aware of a 'shadow' flashing rapidly down his right side. Shortly afterwards the London Mil controller asked the flight to confirm that both aircraft were level at FL110, and at this point the No 2 climbed rapidly to join his leader. Some time later the F-15s were advised that the pilots of G-BYAN had filed an airborne Airprox report

Above:
A Boeing 757-200, similar to G-BYAN, in Britannia Airways livery.
The Aviation Picture Library

— submitted by a pilot or controller if he/she knows or thinks that there has been a loss of safe separation — and only then did the front seat occupant of the No 2 Strike Eagle associate the 'shadow' with the possible presence of another aircraft. The instructor pilot in the rear seat saw nothing of the 757.

*   *   *   *   *

The Daventry Radar Corridor is 8nm wide at FL100, with FL110 available as an alternative level when required. The corridor is in airspace controlled by MTC based at the London Area Terminal Control Centre (LATCC) at West Drayton, to the west of London. Military traffic flying at right angles through the corridor is controlled by London Mil controllers, who are also located at LATCC. A London Mil control cell co-ordinates the use of the corridor by telephone with MTC. Once use of the corridor has been co-ordinated and approved, the civil MTC controllers should regard the corridor level as sterile airspace and give directions to ensure that airliners under their control maintain at least 1,000ft vertical separation from the corridor.

Military aircraft wishing to use the corridor must provide London Military Radar with a minimum of five minutes' notice of their intention to use the corridor. Pre-notification can be given either prior to departure or by radio once airborne. Because of the close proximity of Lakenheath to the eastern entrance of the corridor, there are special corridor procedures laid down for departing F-15s to save time.

If a flight plan has been filed, the information is provided to the London Mil controller on a printed strip. The planned number of aircraft making the crossing is provided on this strip, but military aircraft often go unserviceable at the last minute, which can affect the actual number of aircraft that get airborne. When pre-notification is given verbally over the telephone, the number of aircraft may or may not be mentioned by the operations staff. The system relies on the London Mil controller asking for the number of aircraft from the operations staff.

There was some confusion with the pre-notification of the F-15E flight on 22 November. A flight plan had been filed for a pair of F-15s, callsign EAGLE 31, and a printed flight progress strip showing a flight of two aircraft planning to use the Daventry Radar Corridor had been prepared and was at the London controller's position. At 10.03hrs, Lakenheath Departures called London to pre-notify F-15 traffic for the Daventry Corridor, callsign BOLAR 31. The London controller confirmed that this traffic was in fact the same flight pre-notified as EAGLE 31, and prepared a handwritten flight progress strip reflecting the new callsign. But he assumed that the flight was now a single aircraft and he annotated the handwritten strip accordingly. The printed strip for EAGLE 31 showing a flight of two aircraft was discarded.

At the time of the Airprox, the Midland Terminal Control position in LATCC was manned by a controller and a co-ordinator. The former controlled aircraft directly by radio while the co-ordinator helped plan and co-ordinate aircraft movements and liaised with other agencies by telephone. About 10 minutes before the incident the MTC co-ordinator agreed with the London Mil cell a Daventry Radar Corridor crossing for a 'single F-15' from Lakenheath at

FL100, and provided the MTC controller with the normal flight progress strips giving details of the flight.

The 757's departure for Cyprus would pass through the Daventry Radar Corridor, so initially G-BYAN was cleared to climb to FL90 to give 1,000ft of separation on the Daventry Corridor traffic at FL100. About two minutes before the incident the co-ordinator received a telephone call from London Mil requesting a climb for the corridor traffic to FL110. After checking with the controller, the co-ordinator agreed to the climb and in accordance with normal procedures the MTC controller cleared the 757 to climb to FL100 after he had observed what he thought was one F-15 level at FL110 on SSR Mode C. The MTC controller advised the 757 of the crossing military traffic 1,000ft above; this was acknowledged by the airliner crew who stated that the traffic was indicating on their TCAS. At this stage MTC had not been advised of the second F-15E, and neither the controller nor the co-ordinator had noticed the primary radar return of the rear F-15 that was partly obscured on the radar display by the lead fighter's SSR squawk.

In the 15 minutes before the incident, three different controllers occupied the relevant London Mil control position at LATCC. The first controller received the pre-notification of the F-15 corridor crossing from Lakenheath Operations. The second controller telephoned MTC to co-ordinate the crossing of the Daventry Radar Corridor by a 'single F-15', and subsequently took control of the aircraft from Lakenheath Departures and cleared it to fly through the corridor under its own navigation. At this time the London controller was also arranging an airways crossing for a military aircraft departing from the Royal Naval air station at Yeovilton.

When the F-15s first called on the London Mil frequency, the No 2 aircraft transmitted to indicate to his leader that he was on frequency. This should have told the controller that there were two aircraft in the flight, but transcript recordings indicated that the second aircraft's transmission was partially obscured by the sound of an engaged (busy) tone on an open telephone line to Yeovilton. A third controller took over control of the F-15s shortly before they entered the corridor.

One of the third controller's first acts was to check the flight progress strip of the F-15 and to re-confirm with what she thought was a single F-15 that it was cleared through the corridor at FL100. When the F-15s requested a climb to FL110 the controller noted that descending civilian traffic would prevent an immediate climb, and she told the aircraft to 'Standby, maintain FL100.' The lead F-15 acknowledged the instruction to maintain FL100 and advised that there was a second aircraft in 2km trail.

This was the first point at which any of the London Mil controllers were aware of the second aircraft. However, the controller assumed that her predecessors in the control position had known of the second aircraft, and that there had simply been a mistake in completing the flight progress strip; she amended the strip accordingly. The London Mil controller then contacted MTC to co-ordinate a climb to FL110, but because she assumed MTC was already aware of the second aircraft, she did not mention its presence in 2km trail.

About two minutes before the incident the London Mil controller cleared the F-15s to climb to FL110, and about one minute before the incident she advised the F-15s of civilian traffic climbing 'not above FL100'. The lead F-15 responded that he had the civilian traffic in radar contact.

The civil bible of Air Traffic Services states that clearance for formation flights to enter controlled airspace may be granted provided the aircraft of the formation can maintain separation from each other visually, and all aircraft can communicate with the formation leader. The book of rules goes on to state that all ATC instructions and clearances will be addressed to the leader.

Much of the problem stemmed from confusion over how many Strike Eagles were in the Daventry Corridor. The USAF commonly builds its formation callsign allocation around a single aircraft callsign such as BOLAR 31. Additional elements would simply add consecutive numbers to the prefix; thus, numbers two, three and four in a formation could be BOLAR 32, 33 and 34. However, when dealing with ATC as a single speaking unit, the formation would be known only as BOLAR 31. From an ATC perspective, BOLAR 31 could be a single aircraft or a formation of many.

British military flying regulations state that a formation leader is responsible for separation between the individual units comprising the formation. For separation from other aircraft, formations may be considered as a single unit provided that the formation elements are within one nautical mile both horizontally and longitudinally and are at the same level or altitude. At the controller's discretion, these distances may be increased to 3nm and/or 1,000ft vertically. For stream formations of more than one mile but less than three miles in length, only the lead aircraft is required to squawk Mode 3/A and Mode C. Controllers are to identify the full extent of the stream formation during radar handovers, but there is no requirement for formation elements not in visual contact with the leader to confirm that they have received or acted upon ATC instructions or clearances.

AAIB investigators concluded that this incident resulted from the failure of the pilots in the second F-15E to hear and act on a radio call from London Mil clearing the formation to climb to FL110. Because of inter-cockpit discussion between the two pilots the second F-15 missed the clearance to climb, plus at least two other transmissions that might have made them twig that the climb clearance had been missed. Since there will always be potential for radio calls to be missed, especially in a training environment, the flight safety emphasis was not on blaming anyone for a human error but rather to focus on changing procedures and systems to minimise the effects of a missed clearance by a formation element.

Starting with the Daventry Radar Corridor procedures, the general consensus among civil and military controllers is that radar corridors for military aircraft crossing controlled airspace normally work well. The use of three London Mil controllers in the space of a few minutes was not ideal but is not considered to have had a material effect on the incident. Indeed, with the exception of the confusion over the number of aircraft in the formation, the procedures for feeding the F-15Es across the busy airway system worked well on 22 November.

Moreover, there was no evidence that confusion over aircraft numbers caused this incident. Separation between civil traffic under the control of MTC and military traffic in the corridor is based purely on the requirement for the MTC controller to provide traffic with 1,000ft of vertical separation on the corridor

traffic's cleared level. Even if the MTC controller had been aware of the second aircraft, he would have been entitled to assume that both aircraft were at their cleared level, and by providing the 757 with 1,000ft of vertical separation he complied with the required separation standards.

Turning to the ATC formation procedures, civil air traffic regulations make it clear that air traffic instructions should be issued only to the leader of a formation and the unwritten assumption is that the other members of the formation will either hear the same radio clearance and respond accordingly, or receive instructions separately from the formation leader. If communication with ATC or between formation elements is lost or missed, the official requirement that all formation elements be in visual contact provides a safety backup since formation elements seeing the leader manœuvre in accordance with ATC instructions would be expected to follow the leader. But as the second aircraft was not 'visual' with his leader — the only member of the pair squawking an SSR code — this safety backup was not available. The lack of any requirement for the second aircraft to acknowledge the climb clearance or to call when established at the new level, plus the lack of SSR data available to the controllers, meant that nobody was aware that the No 2 had missed the clearance and had not climbed.

On the face of it, this incident began because of confusion over how many fighters were flying though the

Above:
An F-15E belonging to the 48th Fighter Wing based at Lakenheath in Suffolk.
The Aviation Picture Library

Daventry Corridor. But such things happen in life, especially in aviation where pressures and delays are par for the course. The important thing about the events over Daventry on the morning of 22 November is that what amounted to a misunderstanding should never have got anywhere near the point where 244 bodies and 300,000lb combined weight of metal and fuel could have cascaded down on the East Midlands.

The first mid-air collision between civil airliners occurred on 7 April 1922 when a DH18 and a Farman Goliath collided head-on about 60 miles north of Paris, killing all seven people on board. The answer then was to separate aircraft into air lanes but, once it became clear that more and more aircraft would try to fly more often in finite airspace, the aviation world realised that advances in collision avoidance technology are critical to keep aircraft adequately apart.

In 1993, transport category aircraft in the US began to be equipped with Traffic Alert & Collision Avoidance Systems (TCAS). The requirement for certain categories of civil aircraft either registered in the UK or flying within UK airspace to be equipped with an airborne collision avoidance system took effect on 1 January 2000. This relies on TCAS II, which uses SSR transponder returns to calculate potential airborne conflicts and automatically provides the flight-deck crew with alerting and collision avoidance information. TCAS can provide alerting information on any aircraft transmitting an SSR code, but collision avoidance guidance can only be provided for conflicting aircraft transmitting Mode C or Mode S.

Down on the ground, Short Term Conflict Alert (STCA) is an automated system that alerts controllers to potential conflicts between aircraft returns on the radar display. STCA recognises an aircraft under ATC control by reference to its Mode A code. Conflict alert warnings will only be given for two aircraft where at least one is being controlled from an ATC unit equipped with STCA. When the system detects a potential conflict, flashing SSR labels, suitably coloured to denote the severity of the conflict risk, alert the controller.

All of which sounds marvellous, but aeronautical life is never that simple to regulate. For example, it is difficult to programme the STCA to ignore alerts between aircraft in formation whilst continuing to alert against non-formation traffic. Current ground radar systems and radar displays also have limitations when dealing with aircraft squawking whilst in close formation. SSR labels from aircraft in close formation tend to overlap on radar screens, making it difficult for controllers to read and validate data. Modern radar displays can orientate SSR labels to minimise overlap, but the reorientation procedures on older systems are cumbersome and difficult to use. A further limitation of current radar systems is a phenomenon known as 'garbling', which can occur when data arriving at the SSR sensor from one aircraft overlaps with data from another. Modern monopulse SSR sensors include techniques to minimise the effects of garbling, but there is currently no completely effective degarbling mechanism.

The introduction of automatic safety systems such as TCAS and STCA has generally proved successful. Had they been able to act as advertised on 22 November, what ended as a near-disaster would more likely have been written up as a disconcerting but relatively low risk loss of separation. However, for TCAS and STCA to give of their best it is essential that conflicting aircraft are transmitting SSR information. As it happened, the lack of a squawk from the second F-15E rendered the ATC controllers blind to the developing situation and thus both automatic and 'manual' safety nets were inoperative.

The requirement to carry and operate SSR equipment was introduced to cope with the increasing volume of air traffic and the complexity of ATC procedures, and on current trends the importance of SSR is likely to increase rather than decrease. Given the dependence on SSR of both automatic and 'manual' ATC safety systems, it seems imprudent to operate routine flights in controlled airspace without all aircraft squawking.

As a result of the near miss between the 757 and the F-15s, the AAIB recommended that the British authorities 'should, without delay, implement procedures by which the safety assurance based on the use of SSR is established for aircraft operating in formation'. The CAA accepted this recommendation and the UK Ministry of Defence promulgated a temporary amendment to its flying orders, which introduced wide-ranging changes to formation procedures. The effectiveness of these temporary procedures is to be assessed before permanent changes are implemented. In particular, the MoD reduced the maximum permitted separation between aircraft within formations receiving an ATS service and introduced further restrictions for formations flying in controlled airspace. Revised RT procedures aimed at preventing confusion over the number of aircraft in formation and ensuring that all aircraft are at the formation's assigned level or altitude were also introduced. Civil and military authorities independently but co-operatively commissioned research into SSR garbling and the problems associated with formations and STCA with the aim of allowing all elements of a formation to be allocated individual SSR codes. All these revised procedures should reduce the chances of a recurrence of this type of incident.

It was evident during the investigation that there were differences between military and civil ATC formation procedures. Investigators also noted that some civil controllers were unaware that some formation elements operating in accordance with military regulations would not be in visual contact with the leader. Whilst there is no evidence to suggest that these differences in procedure helped cause this incident, the lack of a good understanding between civil and military controllers can never be in the best interests of flight safety. The CAA has recognised this issue and has requested more information on formation procedures from the MoD with a view to improving the situational awareness of civil air traffic controllers.

Finally, the USAF system of formation callsign allocation has the potential to generate uncertainty over the actual number of aircraft in a formation. One of the interim amendments to British military flying orders requires ATC controllers to ensure that data on the number of aircraft in the formation is obtained before providing a service.

But such procedural and technical refinements can only achieve so much in skies that are becoming ever more crowded with bigger and faster aircraft. Just over two months after the Daventry incident, Japan Airlines (JAL) Flight 907, a Boeing 747-400D, took off from Tokyo's Haneda airport for Okinawa. The 747-400D (Domestic) features a two-man cockpit and the latest in aircraft operating systems. Although designed for long-range flights, JAL had

introduced a number on domestic flights to replace older models and to benefit from lower fuel consumption.

During the climb to an assigned FL390, the crew of the 747-400D saw Flight 958, a McDonnell Douglas DC-10-40 out of Pusan, South Korea, to their front left about 40nm away heading towards Tokyo's Narita airport. The DC-10 was cruising at FL370 and, as the 747-400 approached that altitude, the ATC controller — who was a trainee — told the 747 commander to turn left, into the path of the DC-10. The controller was aware of the possible conflict, but he wanted to bring the DC-10 down under the 747 by setting up its approach into Narita.

Unfortunately, the ATC controller mistakenly told the 747 to descend to FL350 instead of the DC-10. The captain of the 747, which had already climbed through FL370, levelled off to meet the ATC request and he verbally confirmed it. The controller thought that the confirmation came from Flight 958's flight-deck. As the airliners came ever closer, the controller twice asked the DC-10 captain to turn to avoid the 747. The DC-10 captain did not respond, and he said afterwards that none of the three flightdeck crew members recalled hearing the two commands.

About 25 seconds after the 747 was told to descend, each aircraft's TCAS began to show a 'traffic' advisory. At that point, the ATC trainee's supervisor took over and tried to direct the airliners away from each other. But she used an incorrect flight number when talking to the DC-10, and the tri-jet's crew never responded.

The traffic warnings were quickly followed by 'resolution' advisories giving instructions to each pilot on how best to avoid a collision. TCAS advised the 747 captain to climb, but seeing the approaching DC-10, he felt it best to ignore the TCAS advisory and kept descending. In fact he pushed the aircraft's nose down even more. The DC-10's TCAS had shown another aircraft about 13nm away 'at 10-11 o'clock'. The DC-10's captain heeded his TCAS recommendation to descend, but he soon 'decided that the other aircraft was at the same altitude and was descending too, so he stopped the descent'. At around 36,800ft, about 10nm south of the Tokyo suburb of Yaizu, the 747 passed under the DC-10 'at an estimated vertical distance of about 10 meters'.

Initial investigations have tended to point the finger at an ATC trainee's mistake in starting the chain of events that eventually brought two JAL wide-bodies within heart-stopping distance of each other. The near catastrophe was set in chain because the controller, who had three years' experience and was converting to a particular sector, told the wrong airliner to descend and then did not realise his mistake as the situation grew more serious. A supervisor then took over but used the wrong flight number, which further complicated the situation. But what were the two aircraft captains doing, watching the situation develop to the point of near-catastrophe? There could be no excuses about language — two sets of highly paid Japanese aircrew flying JAL aircraft over Japan while talking to Japanese controllers seemed to misread the plot completely.

Left:
A Japan Air Lines DC-10, seen here at Hong Kong's Kai Tak airport, similar to that which had the close encounter with the JAL Boeing 747 near Tokyo.
The Aviation Picture Library

Following the near-collision, the DC-10 carrying 250 people landed safely at Tokyo. The 747, with 427 on board, returned to Tokyo after incurring damage to cabin ceiling panels caused by a drinks cart careening about when the airliner dived to avoid the DC-10. Treatment was necessary for 30 injured crew and passengers. Fortunately, nothing more serious was lost over Tokyo beyond professional pride. But just like the mix-up over Daventry, the Tokyo incident showed that the latest gee-whiz kit in the most modern aircraft such as the F-15E or the 747-400D should never be expected to keep aircraft separate by itself. There will always be a need for sharp human brains on the ground and in the air to keep the skies free from collisions.

Above:
A Japan Air Lines Boeing 747 similar to that involved in the near-miss with a JAL DC-10 close to Tokyo.
The Aviation Picture Library

6

11 September 2001

People who hijack aircraft often claim to be jumping the queue to a better life. In war-ravaged Kabul in January 2000, members of the Young Intellectuals, a secret political group formed to oppose the Taliban, were worried that the security forces were closing in on them. Four of its members were missing, and if caught, the rest of the group could expect to be tortured, beaten and summarily executed.

The leader of the Young Intellectuals was Ali Safi, 38, a former soldier and university lecturer who had been arrested and held by the religious police for a week just for playing chess. He hatched a plot to hijack an airliner from Kabul Airport together with his brother, Mohammed, 33. As the group's 'military director', Mohammed was able to obtain three Mausers and one other automatic handgun, two hand grenades and two impact detonators. Tickets, costing £12.50 each, were bought for the group, totalling 58, including some family members. The majority were told they were fleeing the Taliban, but not how or when.

Just before dawn on 6 February 2000 they gathered at the bombed-out shell known as Kabul Airport. Under the cover of going to a wedding, they underwent only cursory checks of their belongings. Ten minutes into what was supposed to be a 40 minute flight to the northern city of Mazar-i-Sharif, the men got up from their seats, brandishing pistols, hand grenades and knives which had been smuggled on board by officials who had been bribed.

The hijackers took control of the Boeing 727, owned by Afghanistan's national airline, Ariana. They screamed at the passengers not to move. A man suspected of being a Taliban member was tied up along with members of the cabin crew, while the pilot was ordered at gunpoint to fly over the country's western mountains. It was the beginning of a journey that would see the 727 land in Tashkent in Uzbekistan, where the authorities initially refused to give them fuel because they did not have any money to pay for it, then a forced landing in Kazakhstan due to engine failure, and a tense few hours on the ground at Moscow airport where the airliner was surrounded by heavily armed soldiers.

The men had initially planned to fly to Switzerland but then Safi persuaded the pilots to fly to Britain. They finally landed at Stansted, Essex, at 02.01hrs on 7 February. By now the cabin was in a terrible state — the toilets were full and the air conditioning was doing little to cleanse the fetid air. But the hijackers were in no hurry to get off. Stansted was the designated British airport for dealing with hijackings and Essex Police, who were well versed in what was required, used the tried and tested tactic of talking the men into submission. Contact was made with the hijackers over the radio and every request they made for food, water and medicine was endlessly negotiated. By the time the drama ended, the hijackers' chief negotiator had lost his voice.

The police soon realised that they were not dealing with a band of hardened terrorists. This was confirmed late on the second night when a rope suddenly appeared from the cockpit, closely followed by four members of the aircrew. The hijackers had allowed their most precious hostages — the pilots who could fly the 727 — to escape.

It was more than an hour before the hijackers realised what had happened. Then bedlam ensued. They broke down the cockpit door with an axe and the remaining crew were trussed and beaten as they lay on the aircraft floor. The hijackers threatened to kill them, and all the passengers on board, unless the escapees returned. The stewards begged Police negotiators to do something before the radio abruptly went dead.

A steward, who suffered a head wound in the attack, was shoved down the steps of the airliner with his hands tied behind his back. For both Police and passengers this was the most dangerous period of the hijacking. The marksmen and SAS team surrounding the 727 were ready to storm the cabin if the hijackers began shooting, but as the night wore on tempers cooled. The hijackers, who had already released more than 30 hostages, including a man who had become ill and a number of women and children, prior to reaching Stansted, continued to allow a steady drip of passengers off the aircraft.

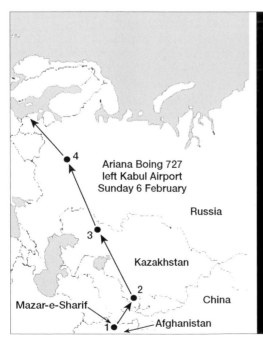

Ariana Boing 727
left Kabul Airport
Sunday 6 February

Stansted Hijack

1. Ariana flight from Afghanistan capital Kabul is hijacked en route to Mazar-e-Sharif.

2. Tashkent, Uzbekistan. Aircraft lands to refuel. 10 passengers are allowed off.

3. Aktyubinsk, Kazakhstan. Leak in right fuel tank repaired.

4. Moscow, Nine allowed off before aircraft leaves for Stansted.

5. Stansted. Aircraft touches down from Moscow with 150 people on board. It is isolated away from the terminal buildings on the south side of the airport.

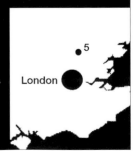

Over 90 hours after the flight had left Kabul, two of the hijackers left the 727 to talk directly with a Home Office negotiator and a representative of the UN High Commissioner for Refugees in the UK. At 03.17hrs the remaining hostages began to leave the aircraft, followed less than two hours later by the hijackers themselves. It was just after 05.30 on 10 February that Britain's longest hijack drama ended.

Nine Afghan men who hijacked the Boeing 727 stood trial in the Old Bailey. The jury found them guilty by a 10 to one majority of hijack, and unanimously of false imprisonment of passengers and crew and possessing weapons. The two Afghan brothers who masterminded the operation were jailed for five years. Six of their followers were jailed for 30 months, and a seventh for 27 months because he was only 18 at the time of the hijacking. The judge, Sir Edwin Jowitt, accepted that the Young Intellectuals of Afghanistan had initially been fleeing the Taliban regime. However, he said their actions had turned into a criminal act after the Ariana 727, with its 164 passengers, was forced to fly on to Britain after landing in Moscow. Sir Edwin said the brothers had prolonged the 70-hour siege at Stansted airport in Essex in order to make a political point.

Notwithstanding the post-action justifications, all aircraft hijackings apart from those committed by the mentally unstable are undertaken to make a political point. On 24 December 1994 an Airbus A300 (F-GBEC) belonging to Air France was scheduled to fly from Houari Boumediene Airport, Algiers, for Paris Orly. As Flight 8969 was being prepared for departure, the crew of 12 and 237 passengers were hijacked by terrorists from the Groupe Islamique Armée, who were protesting against French political, military and economical support for the Algerian government. Two passengers were shot after about nine hours, followed by another on Christmas Day. On 26 December F-GBEC was flown to Marseilles where it was stormed by security forces around 17.00hrs. All four hijackers were shot and substantial damage was caused to the cockpit. The hijackers' conversations had been monitored, and a robust French response was assured when the hijackers revealed their intention to get airborne and fly F-GBEC into the Eiffel Tower.

Evil ideas appeal to evil minds and on 11 September 2001 the US became the target of the largest terrorist incident in history when Osama bin Laden's al-Qaida organisation hijacked four separate US airliners and used them as jet-powered missiles to attack distinctive symbols of American society in New York City and Washington DC.

At 05.30hrs on Tuesday 11 September, two young Arabs — an Egyptian called Mohammed Atta and a Saudi, Abdulaziz al-Omari — left the Comfort Inn in Portland, Maine, without breakfast. Twelve minutes later they went through security checks at Portland Jetport before boarding Flight US5930 for the hop to Boston. A security camera recorded Atta striding purposefully forward — at Boston he had another plane to catch. In Boston, Newark and Washington, another 17 young Arabs were also dressing for the last time.

The plan to fly into American buildings was not Atta's invention. The concept went back to at least the Air France hijacking in Marseille, and the outline was given to Atta as leader of the attack to develop in detail. He found ready accomplices. One was a Lebanese

Left:
The route flown by the hijacked Ariana 727 from Kabul, Afghanistan, to Stansted, Essex.

student, Ziad Jarrah, who had known nothing but civil war during his early life in faction-torn Beirut and the Bekaa Valley. He moved to the east German town of Greifswald to study aeronautical engineering, and his ambition was to be a pilot. By the summer of 1997 his German was almost fluent and he won a place at the same technical university in Hamburg where Atta was studying.

Another bright student called Marwan al-Shehhi had been born in the United Arab Emirates. His father was the local imam, he married a local girl and he attended university in Germany where he was known for his sense of humour and Benetton shirts. It was also in Hamburg, where he went to study ship-building, that he met Atta. Al-Shehhi's background was hardly the conventional stuff of poverty and deprivation, but notwithstanding their disparate back-grounds the trio of Atta, Jarrah and al-Shehhi would seize their 15 minutes of fame and the controls of three airliners on 11 September.

In late 1999 Mohammed Atta took his key lieutenants to Afghanistan where they refined their conspiracy. Among those they recruited, largely to provide the 'muscle' on the suicide hijackings, no fewer than 11 came from the scenic highlands of Asir in the west of Saudi Arabia. Some of them had excellent prospects but, for whatever reason, by the time these young men arrived in the USA they had all agreed to die for their cause.

After Atta and al-Shehhi arrived in the US in July 2000, they each handed over a $10,000 cheque for a five-month flying course. Instructional staff at Huffman Aviation in Venice, Florida, were not impressed with the arrogant, inattentive and rather stupid students, but by the end of 2000 Atta and al-Shehhi had their private pilot's licences. This was the most basic qualification for small, single-engined aircraft, and we now know that Atta moved on for further training at Palm Beach Flight Training in Coral Springs, Florida.

It is estimated that Atta's people went through almost $90,000 during the run-up to 11 September, all of it wired from a bank account in Dubai by one of bin Laden's most trusted moneymen. The third member of the Hamburg ring, Ziad Jarrah, then arrived from Germany. Jarrah joined the US1 Fitness Center in Dania Beach. The owner recalled that Jarrah 'wanted to learn about fighting and control — about being in control and how to control somebody'.

Atta spent his last hours in Portland and al-Shehhi stayed in Boston. The fourth team stayed in Maryland close to Dulles Airport. Their leader was Hani Hanjour, who had entered the US on a student visa but had never taken up his studies. Hanjour was a trained commercial airline pilot and the son of a wealthy businessman from Ti'af, 80km east of Mecca. One member of his team was Majed Moqed, a law student from the King Saud University in Riyadh, and the son of a head of the Baniauf tribe from a village near Medina. Moqed was known for his desire to join a jihad and he was a relative of an extremist leader who took part in the 1979 storming of Mecca's Grand Mosque, Islam's most holy shrine.

Jarrah's team stayed in the Newark Airport Marriott hotel, just two minutes from the departure terminal they intended to use the following morning. Jarrah paid an extravagant $450 in cash for two non-smoking rooms, each with a clear view of the Manhattan skyline dominated by the twin towers of the World Trade Center. On top of the twin towers, as on all very tall structures, were pulsing red lights put there by FAA mandate to warn aircraft to keep clear.

*   *   *   *   *

Capt John Ogonowski, aged 53 and an air force veteran of the Vietnam War, was going through the pre-flight checks on Boeing 767-200ER N334AA with his co-pilot, First Officer Tom McGuiness. Their first leg of the day was to take American Airlines Flight 011 from Boston Logan to Los Angeles International Airport. Passengers were checking in and, as usual, security was lax. An X-ray scanner for bags and a metal detector stood at the entrance to one wide corridor leading to the departure gates. The equipment took up only half the corridor. The other half was given over to disembarking passengers, and people could walk back and forth without being checked. Between 1997 and 1999 the FAA had fined the Massachusetts Port Authority, which ran Logan, $178,000 for 136 incidents of failing to screen baggage or failing to restrict access to designated 'secure areas' of the airport. Small wonder that five Middle Eastern hijackers, armed with plastic razor-blade knives, aroused no suspicion as they passed through 'security'. Mohammed Atta was booked into seat 8D in business class. Another Arab, Abdulrahman Al-Omari, was seated next to him and three more associates were also on board. They all left the ground at 07.58hrs.

Five further hijackers led by al-Shehhi were boarding a second 767-200 (N612UA) at Logan. This was United

Left:
N334AA, the American Airlines Boeing 767-200 that was the first aircraft to be flown into the World Trade Center on 11 September 2001. *The Aviation Picture Library*

Airlines Flight 175 to Los Angeles and it got airborne just after American 011. At 08.01hrs, United Flight 093, a Boeing 757-200 with Ziad Jarrah's team on board, departed Newark bound for San Francisco International. Another 757-200, American Airlines Flight 077, left Dulles Airport, Washington DC, at 08.10hrs for Los Angeles with Hani Hanjour plus four on the passenger manifest. All four twin-engined 757s and 767s were fully fuelled for their six-hour flights.

The two 767s headed northwest from Boston initially and they came close to colliding at one point, which indicated that hijackers had moved swiftly. On Capt Ogonowski's American Flight 011, businessman Peter Hanson rang his family in Connecticut and said that a flight attendant had been stabbed. After passengers were herded to the rear, the hijackers burst on to the flightdeck. ATC overheard a brief snatch of conversation along the lines of, 'Don't do anything foolish. You're not going to get hurt. We have more planes, we have other planes.' Nothing more was heard from Flight 011. At 08.28, area radar controllers saw the 767 make a sharp turn south.

United 175 flew normally for longer, and the hijackers appeared to have left one authorised pilot in his seat. A flight attendant made a distress call on a special line to United's flight operations centre reporting that passengers were being stabbed. She gave the seat number of a hijacker around the time this 767 also turned sharply south. Marwan al-Shehhi on the flightdeck had learned enough in his training to turn off the transponder that enabled ground controllers to pinpoint the airliner's altitude and position. From now on, watchers on the ground would have to rely on raw radar returns. These were used to vector two F-15 interceptors scrambled from Otis Air Force Base at 08.39hrs, but no one on the ground understood what was happening or what should be done. Even if they had put in full afterburners, the F-15 pilots could have done little because the first airliner was just six minutes' flying time from Manhattan.

Up to 11 September, the USA had relied on an awesome shield of missiles and similarly sophisticated weaponry to deter any attack on the American homeland. It was a pretty audacious plan by bin Laden's al-Qaida network to hijack four 300,000lb 'fuel-air bombs' and, out of eastern seaboard skies busy with commuter airliners, to launch multiple attacks on multiple targets from multiple axes.

Inside the north tower of the World Trade Center, Anne Prosser was riding the lift to the 90th floor. It was 08.46 on another blue-sky day and as the lift doors opened 1,000ft above the city known as the Big Apple, an explosion blew Anne out of the lift and sent her sprawling. A few floors above her, 150 tonnes of 767 carrying 10,000gal of high-octane jet fuel had slammed into the north tower between the 96th and 103rd floors at more than 300mph. From his loft room, British designer Colin Heywood saw 'a plane tilt and head towards the building. As it hit, its wings seemed as wide as the tower. It disappeared into the building, as if it was swallowed inside.' The flight crew of 11 and 81 passengers stood no chance.

At first everyone thought there had been a terrible accident. In the south tower, a financial broker heard a security guard announce, 'This is Tower Two. A plane has gone into Tower One. Two appears secure. You can go back to your offices if you want.' He was not to know that al-Shehhi would aim United 175 at the middle of the south tower between the 86th and 97th floors. He could hardly have missed such a conspicuous target, and nine United crewmembers and 55 other passengers died with him at 09.03hrs.

Fifteen minutes later, American Flight 077 bound for Los Angeles at FL350 turned east and was lost to radar. Hani Hanjour was now in control of 757 N644AA and Barbara Olsen, a former congressional aide and biographer of Hillary Clinton, was locked in the lavatory calling her husband, Solicitor-General Ted Olsen, at his office in the Justice Department. She told him they had 'knives and cardboard box cutters' which they had brandished to round up the passengers at the rear of the 757. She asked what she should tell the pilot to do, but 51-year-old Capt Charles Burlingame was probably dead.

As the magnitude of what was happening became clear, the Vice-President of the United States was hurried into a bombproof bunker and other staff were evacuated. President Bush was 850 miles away in Texas, and he was probably unaware of the full horrors unfolding on the hijacked airliners as the third flying fuel bomb descended towards the nation's capital from the north. It appeared to be heading for the White House but then Hanjour appeared to swing the 757 right in a broad turn through 270°. He levelled the wings, applied power and screamed low over the Columbia turnpike, a busy commuter slip road. Witnesses could only watch dumbfounded as the 757 with its 58 passengers and six crew smashed across the Pentagon's helipad, swept aside a fire engine, demolished a military helicopter and buried itself into the 2ft-thick west wall of the Pentagon. The 757 disappeared completely — no fuselage, no tail and no wings were visible. It tore a 100ft-wide hole in the Pentagon, ploughing through three of the building's five concentric rings of corridors and blasting holes through the last two rings to the main inner courtyard. Fires spread 300m either side and were still burning three days later. It has to be said that Hanjour piloted the kamikaze 757 with great precision. A degree too low and he would have driven into the ground first; a degree too high and he would have overshot and landed in the Potomac. The attack missed the offices of the Defense Secretary and the most senior military officers but that did not matter. From the al-Qaida perspective, at 09.43 they had plunged a stake into the heart of the US military machine that had offended so many Arabs in recent years.

That left United Airlines Flight 093 and by now all hell had broken loose on the 757 (N591UA). Capt Jason Dahl, a 43-year-old who had flown with United since 1993, was heard on the open cockpit microphone telling someone to 'get out of here' and then a foreign voice warned that a bomb was on board. United 093 was seen to turn south off its route to San Francisco while over Cleveland at FL350, by which time an hour had passed since the first 767 struck the World Trade Center. As the 38 passengers and seven crew on United 093 learned of the earlier suicide attacks via their mobiles, some determined that they had nothing to lose by tackling the hijackers. Cellphone messages well into the two-hour flight indicated that both pilots were under guard at the rear of the aircraft with cabin crew and passengers. Given that the majority of the subsequent aircraft wreckage would be confined to a 100m-radius, it seems that this group grappled in hand-to-hand contest with the hijackers rather than a bomb went off. The important thing was that United 093 was not allowed to hit whatever Ziad Jarrah intended to fly into. Down below, Michael Merringer on his mountain bike heard the Pratt &

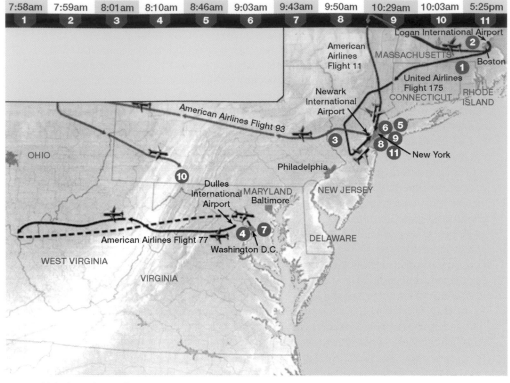

7:58am	7:59am	8:01am	8:10am	8:46am	9:03am	9:43am	9:50am	10:29am	10:03am	5:25pm
1	2	3	4	5	6	7	8	9	10	11

1. United 175 departs Boston
2. American 011 departs Boston
3. United 93 departs Newark
4. American 77 Departs Washington
5. American 011 hits the north tower of the WTC
6. United 175 hits the south tower
7. American 77 flies into the Pentagon
8. The south tower of the WTC collapses
9. The north tower collapses
10. United 93 crashes in Pennsylvania
11. A third WTC building collapses

Whitney engines surge and saw the 757 wobbling. At this stage N591UA was out of any one man's control and it eventually came down at 10.03hrs in a field in Somerset County, Pennsylvania, 130km southeast of Pittsburgh. Given that Jarrah was heading towards Washington when control was wrested from him, and he would have been going for something both symbolic and distinctive, he was probably aiming for the White House.

At this stage 266 passengers and crew on four hijacked airliners had died. Apart from 100-plus workers who had died in the Pentagon, casualties on the ground had been miraculously few. The 1,350ft towers of the World Trade Center had been built to withstand the impact of a Boeing 707, and when the first 767 smashed into the top of the north tower, the building just shook and swayed. But, just after Hanjour ploughed into the Pentagon, there was a terrible sucking noise at the twin towers. A cloud of dust billowed out of the south tower's top storeys like some grotesque roman candle. The intense heat generated by the huge flying fuel bombs melted the steel frames that held the towering structures upright.

As two or three floors melted away, the floors above 'pancaked' down on each other, entombing all inside. Everything underneath them came down like a pack of cards because of the huge weight. First, the south tower went down around 09.50, followed by the north tower at 10.29. A third building in the World Trade Center complex collapsed at 17.25 that afternoon. Around 3,000 firefighters, police, office workers and others unfortunate enough to be just minding their own business in the wrong place perished as a consequence. As 2,400 lives were lost at Pearl Harbor in 1941, the hijackings on 11 September resulted in the largest single-day loss of life on US soil since the Civil War.

Fifteen of the 19 hijackers are now known to have been Saudis, including the brother of a police commander, the son of a tribal chief, two teachers and three law graduates.

There is no obvious reason why young men from comfortable backgrounds in one of the world's richest nations felt moved to kill in cold blood and follow Mohammed Atta's urging to 'welcome death for the sake of God'. Mr Atta and three others were competent pilots and two more had spent a few hours learning to fly in San Diego, but this enlightening experience did not appear to lift them above the surly bonds of rage.

Given that investigators are increasingly convinced that one or two other hijackings were in the works, the question for regulators, airlines and passengers is how to prevent such catastrophes in future? Stopping flying would be rather drastic, first because there is no going back to horse and buggy and, more importantly, the terrorists would have won.

To improve flightdeck security, Japan's Transport Ministry believes it is urgent that Japanese airlines introduce physical barriers to the cockpit. Airlines around the world have begun to evaluate enhanced transponder and cabin video surveillance systems. The transponder would transmit continuously in an emergency, and closed circuit video cameras would enable the flight crew to monitor areas outside the cockpit. But all the talk of barred cockpit doors, pilots with guns, plainclothes air marshals on every trip and a future where ground controllers take over hijacked aircraft via datalinks is not the answer.

What will restore and retain public confidence in air travel is to keep potential hijackers and their weapons off airliners. This will flow from greater international co-operation in identifying and tracking prospective troublemakers, using new software to flag up unusual behaviour, such as the purchase of one-way tickets with cash, and developing technology for detecting non-metallic weapons, lethal gases or powders and explosives.

All in all, there must be a shift in security thinking within airports and airlines. 'It's better to be safe than sorry' is the mantra of national aviation authorities, but it is hard to believe that much serious thought has gone into a system that stops a pair of dinky makeup tweezers from getting on board while letting through a seriously sharp fountain pen. We all go through the rigmarole of assuring the questioner that we have packed our own bags and not let them out of our sight, but what is the point? Is anyone seriously going to reply that they gave their bags to the bin Laden packing service? Even if you assume that only half of

Left:
Flight paths of
destruction on
11 September 2001.

87

the 650 million people who board a US domestic flight each year consider the questions for just 15 seconds apiece, by the end of a year that adds up to about 81 million minutes, or 154 years — the equivalent of two lifetimes sacrificed annually to regulations that have never been shown to save a single life. Airport security needs to be better focused, and authorities could start by beefing up hitherto cursory background checks to remedy the situation whereby 80% of the cleaners at one airport were found to be illegal immigrants and two were former convicts. Just after 11 September, one screener at Oakland International Airport said that he earned $7 an hour manning the X-ray checkpoint — and more than $9 an hour at his second job, cleaning tables at a restaurant.

Some put their faith in passenger profiling. Differentiating by ethnic group, religious belief or anything of that nature is not impossible, but it is fraught with legal and human rights issues. It is also of limited utility. Many of the 19 hijackers on 11 September were well educated, intelligent and in their mid-twenties, with a history of making trips both to America and Afghanistan. A leading moderate Saudi dissident observed that 'Whatever you may think of what they did, they were talented people. They were not only credible but had good communication skills.' In other words, no passenger profiling system acceptable in a Western democracy could have picked them out.

Airports hold the security key, and the authorities should pay far greater attention to the airside of the terminal, where caterers, ground staff and anyone looking purposeful in a yellow jacket and holding a clipboard can apparently come and go. Exactly five months after the events of 11 September, robbers overpowered a British Airways driver in a supposedly secure airside area at Heathrow and escaped with the equivalent of £4.6 million in US dollars using a BA vehicle. So much for the tougher security measures introduced after 11 September. While it is far easier to yearn for high-tech fixes, real security lies with recruiting, vetting, training and paying the right people to safeguard airports and those who pass through them properly.

One good thing to come out of 11 September was the realisation that hijacked passengers will no longer sit passively while hijackers take them to Marseille, Stansted or New York. The response of many brave people on United 093 over Pennsylvania should warn any future hijackers that they may now be up against irate and fighting-mad passengers outnumbering them by perhaps 30 or 40 to one. Fire and oxygen bottles are heavy clubs, while seat cushions and lifejackets can be used defensively against knives. Events of 11 September were a wake-up call to all air travellers that security, like flight safety in general, is everyone's business.

Left:
A United Airlines' 757-222 similar to that hijacked on 11 September 2001. *The Aviation Picture Library*

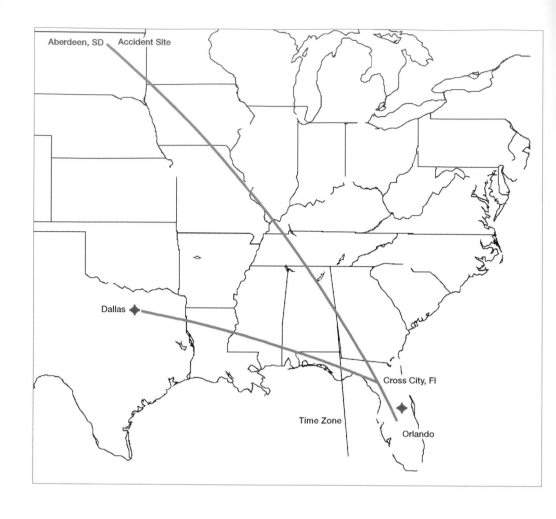

Aberdeen, SD   Accident Site

Dallas

Cross City, Fl

Time Zone

Orlando

Above:
The route flown
by Payne Stewart's
Learjet compared
to the planned
leg to Dallas.

7
Oxygen Failure

William Power Lear was born in the same Missouri town as the writer Mark Twain. At the age of 16, Lear joined the US Navy where he learned radio electronics to such good effect that his development of a practical car radio launched the Motorola Company, and he went on to design the eight-track player in the 1960s. After World War 1 he took up flying. He began designing aircraft navigational aids in the 1930s and his companies filled more than $100 million of defence orders during World War 2. After 1945 he developed the 'Learmatic' lightweight autopilot system, and in 1962 he formed Learjet, which became the leading supplier of corporate jets within five years. This remarkable man died in 1978. He was said to have named his daughters Gonda and Chanda.

The Learjet philosophy was to provide the business world with an executive twinjet capable of airliner performance and comfort. Out of the basic Learjet 25 grew the slightly larger, turbofan-powered Learjet 35, which could sweep along at a cruising speed in excess of 400kt up to 45,000ft over a range of 2,289nm. The Learjet 35 could carry up to eight passengers and this mini-airliner capable of fast jet performance proved so popular that nearly 700 were built between 1974 and the mid-1990s. They retailed at about £3.5 million each.

The rich and famous loved the convenience and comfort of the Learjet, with its leather seats, polished wooden fittings and a mini-bar. Instead of the hassle of getting to a major airport, checking in with the masses and getting held behind a host of other airliners at the holding point or in the stack, those in a hurry had only to get to the nearest municipal airport and they were on their way. For those like pop stars and sports personalities, who had to dash from event to event because time was money, the fast executive jet was a godsend. Learjets, owned by celebrities such as the Hollywood actors Arnold Schwarzenegger and Patrick Swayze, cost up to £1,500 an hour to hire.

The golfer Payne Stewart, like William Lear a native of Missouri, was one of the top competitors on the circuit. His plus-fours, or knickers in American, and his tam o'shanter cap made Payne Stewart one of the most recognisable players in golf. His game and his passion made him also one of the most respected. He played on three winning Ryder Cup teams, and he climaxed an unforgettable day in June 1999 when he sank the longest putt ever made on the final hole to win his second victory in the US Open.

By 1999 Payne Stewart was ranked No 8 in the world and was third on that year's money list, with just over $2 million. On 25 October 1999 the 42-year-old Stewart was to go to Texas, first for a meeting on a proposed golf course near Dallas, then on to the Tour Championship in Houston. The best way to meet this tight schedule was to fly by executive jet.

Learjet Model 35 N47BA was operated by Sunjet Aviation of Sanford, Florida. On 25 October, N47BA's flight crew were scheduled to begin a two-day trip sequence of five flights. The trips on the first day were to be from Sanford International Airport (SFB) to Orlando International Airport (MCO), from MCO to Dallas Love Airport (DAL), Dallas, Texas, and from DAL to William P. Hobby Airport, Houston. The captain reported for duty at SFB about 06.30 (Eastern Daylight Time), and the first officer arrived about 15 minutes later. Both pilots were in a good mood and appeared to be in good health. The first flight of the day was to be a visual flight rules (VFR) positioning flight from SFB to MCO, approximately 15nm away.

The captain asked a Sunjet Aviation line technician to pull the aircraft out of the hangar, fuel it to 5,300lb fuel weight, connect a ground power unit to the aircraft, and put a snack basket and cooler onboard. The first officer arrived at the aircraft just before the fuelling process started and stayed in the cockpit during refuelling. He then went inside the terminal building while the captain made a pre-flight inspection of the aircraft.

About 07.25hrs, an instrument flight rules (IFR) flight plan was filed for the second flight of the day, MCO to DAL, indicating that N47BA would depart MCO about 09.00, follow a route over Cross City, Florida, and proceed directly to DAL. The requested altitude was 39,000ft. The flight plan indicated that there would be five persons on board (two pilots and three passengers), and 4 hours 45 minutes of fuel.

N47BA departed SFB about 07.54hrs, arriving at MCO at about 08.10. The captain told a ground technician that they were picking up passengers and did not require additional fuel. The passengers — Payne Stewart, his agents Robert Fraley and Van Ardan and a last-minute addition — the golf course designer Bruce Borland — arrived about 30 minutes later and boarded the aircraft. Several bags were placed on board the aircraft, including a big golf bag weighing about 30lb. The flight departed MCO at 09.19 bound for DAL. Two minutes later, the flight contacted the Jacksonville ARTCC and reported climbing through 9,500ft for 14,000ft.

The controller instructed N47BA to climb and maintain FL260. N47BA acknowledged this clearance by stating, 'two six zero, bravo alpha'. At 09.23:16 the controller cleared N47BA direct to Cross City and then direct to DAL. N47BA acknowledged the clearance. At 09.26:48 EDT N47BA was issued instructions to change radio frequency and contact another Jacksonville ARTCC controller. N47BA acknowledged the frequency change.

At 09.27:10 N47BA told the Jacksonville controller that the flight was climbing through FL230. At 09.27:13 Jacksonville ARTCC instructed N47BA to climb and maintain FL390. Five seconds later, N47BA acknowledged the clearance by stating, 'three nine zero, bravo alpha'. This was the last known radio transmission from the aircraft. No sound of anything untoward was heard on the ATC recording of this transmission.

At 09.33:38 EDT (6 minutes and 20 seconds after N47BA acknowledged the previous clearance), the controller instructed N47BA to change radio frequencies and contact another Jacksonville ARTCC controller. The controller received no response from N47BA. The controller called the flight five more times over the next four and a half minutes but to no avail.

Staff Sgt James Hicks was on duty as an air traffic controller at Eglin Air Force Base, Florida, when he received a call from Jacksonville Center alerting him that the Learjet might be in trouble. USAF Capt Chris Hamilton, an F-16 test pilot from the 40th Flight Test Squadron at Eglin AFB, was engaged in air-to-air combat training against an A-10 Thunderbolt when Hicks directed him to an airborne tanker operating over Eglin's vast military reservation. After topping up, Hamilton gave chase in his Fighting Falcon, but as the Learjet was a high-performance jet with a 100-mile start, it took time for Hamilton to close with N47BA. At 09.52 Central Daylight Time, over an hour after the last transmission from N47BA, Hamilton caught up with the Learjet over Memphis, Tennessee.

At about 09.54, at a range of 2,000ft from the Learjet and at an altitude of about 46,400ft, the 32-year-old test pilot made two radio calls to N47BA but received no response. At about 10.00 Hamilton began a visual inspection of N47BA. There was no visible damage to the aircraft, and he did not see ice accumulation on the exterior of the aircraft. Both engines were running, and the rotating beacon was on. He could not see inside the passenger section of the aircraft because the windows seemed to be dark. Furthermore, Hamilton saw that the entire right cockpit windshield was opaque, as if condensation or ice covered the inside. The left cockpit windshield was also opaque, although several sections of the centre of the windshield seemed to be only thinly covered by condensation or ice. A small rectangular section of the windshield was clear. Capt Hamilton did not see any flight control movement and at about 10.12 he finished inspecting N47BA and flew to Scott AFB, Illinois. 'It's a very helpless feeling to pull up alongside another aircraft and realise the people inside potentially are unconscious or in some other way incapacitated,' Hamilton said afterwards. 'And there was nothing I could do physically from my aircraft, even though I'm 50 to 100ft away, to help them at all. That was very disheartening.'

At about 11.13 two Oklahoma Air National Guard F-16s — TULSA 13 flight — were vectored to intercept the Learjet by Minneapolis ARTCC. The TULSA 13 lead pilot reported to the Minneapolis controller that he could not see any movement in the cockpit. At about 11.25 TULSA 13 lead reported that the windshield was dark and that he could not tell if it was iced. Eight minutes later, a TULSA 13 jet manoeuvred in front of the Learjet, and the pilot reported, 'We're not seeing anything inside, could be just a dark cockpit though . . . He is not reacting, moving or anything like that. He should be able to have seen us by now.' TULSA 13 were looking at an aerial Marie Celeste.

At 11.39 TULSA 13 flight departed for their in-flight refuelling tanker. Eleven minutes later, while radar controllers routed air traffic around the Learjet and kept aircraft from flying underneath it in case it crashed, two North Dakota Air National Guard F-16s with the identification NODAK 32 flight were vectored to intercept N47BA. TULSA 13 flight then returned after being topped up, and both pairs manoeuvred in close proximity to N47BA. At about 11.57 the TULSA 13

lead pilot reported, 'We've got two visuals on it. It's looking like the cockpit window is iced over and there's no displacement in any of the control surfaces as far as the ailerons or trims.' The only sound inside that Learjet would have been from Mr Stewart's wife, Tracey, who after initial TV reports that the aircraft was flying out of control kept trying to contact her husband on the mobile phone.

At 12.10:41 the sound of an engine winding down, followed by sounds similar to a stick-shaker and autopilot disconnect, were captured on N47BA's cockpit voice recorder (CVR). The CVR also captured the continuous activation of the cabin altitude aural warning. At 12.11:01 ATC radar saw N47BA begin a right turn and descend. A TULSA 13 aircraft followed N47BA down. At 12.11:26 the NODAK 32 lead pilot reported, 'The target is descending and he is doing multiple aileron rolls, looks like he's out of control . . . in a severe descent, request an emergency descent to follow target.' N47BA entered a descending spiral and impacted in an open, flat grassy field near Aberdeen, South Dakota. 'The plane had pretty much nosed straight into the ground,' said a witness who lived two miles from the crash site. The force of the impact was such that the Learjet made a crater 8ft 6in down at its deepest point.

The main airframe wreckage was located in or near the impact crater. The majority of the rest of the wreckage was found within an approximately 75ft radius, although additional wreckage was recovered up to 150ft away. A debris field of smaller wreckage, including instrument panel components, the flight manual, seat cushions, life vests and personal effects, extended outward from the impact crater in a north-northeasterly direction toward a nearby marsh. All on board — Capt Michael Kling, 43, First Officer Stephanie Bellegarrigue, 27, and the four passengers — were killed. There was no fire, indicating that Payne Stewart's Learjet had run out of fuel.

Afterwards, a Pentagon spokesman said that consideration was never given to shooting down the Learjet.

*   *   *   *   *

Although six National Transportation Safety Board investigators were making a preliminary examination of the crash site that Monday night, the investigation into the loss of N47BA took 13 months to complete. It was hampered by several factors. First, because the aircraft impacted at nearly supersonic speed and at an extremely steep angle, none of its significant components remained intact. Consequently, NTSB investigators had painstakingly to examine the fragmented valves, connectors and portions of other aircraft parts before they could draw any conclusions about the accident's cause. Second, the Learjet was not equipped with a flight data recorder, and it had only a 30-minute cockpit voice recorder, which proved to be of limited use during the investigation. Finally, all the NTSB investigators assigned to the case were also investigating other accidents. The Investigator-in-Charge, Bob Benzon, was working on four other investigations in addition to this one.

None the less, the NTSB team did a thorough job. They found that both pilots were properly certificated under company and FAA requirements. Capt Kling had a USAF four-jet background and had been a military instructor-pilot. According to Sunjet Aviation, he had converted to the Learjet without difficulty, he was knowledgeable about the aircraft and he was a confident pilot with good situational awareness. Family and co-workers indicated that he was a non-smoker in excellent health, who did not take medication or consume alcohol.

Pilots who had flown with the first officer indicated that she had good aircraft handling skills. She was known as a serious and knowledgeable aviator who had a 'meticulous' style in the cockpit and who did not abbreviate procedures or neglect checklists. She had excellent radio communication skills and was also a non-smoker in good health. Neither pilot had any record of being involved in previous aircraft accidents, incidents or enforcement actions. Overall, the NTSB concluded that the pilots' duty time, flight time, rest time and off-duty activity patterns did not indicate any pre-existing medical, behavioural or physiological factors that might have affected their performance on the day of the accident. Visual meteorological conditions prevailed along the route of flight, and weather was not a factor in the accident.

Left:
The remains of the Learjet being removed from the cratered crash site in South Dakota.
Associated Press

Turning to the aircraft, the unfortunate Learjet 35 was manufactured in 1976 and had been maintained and operated by Sunjet Aviation since January 1999. The aircraft was properly certificated and equipped in accordance with Federal regulations and approved procedures. No significant pre-existing airframe or turbofan engine problems were discovered, and there was no evidence in the wreckage of an in-flight fire.

The flight crew's last communication with ATC had been at 09.27:18 EDT, when the first officer acknowledged a clearance to flight level (FL) 390 while the Learjet was climbing through 23,200ft. The aircraft was passing over Gainesville, Florida, and the first officer's speech was normal, her phraseology was accurate and appropriate, and NTSB testing indicated that she was not using an oxygen mask microphone for this transmission or those that she had made earlier.

This is an important point. The human body and brain need oxygen to function properly. If you climb up a mountain, breathing gets harder and mental faculties are impaired not because there is too little oxygen but because reduced air pressure prevents enough oxygen from being forced across lung tissues into the bloodstream. Pressurised cabins of turbine-powered aircraft are designed to maintain a consistent cabin altitude of approximately 8,000ft by directing engine bleed air into the cabin while simultaneously regulating the flow of air out of the cabin. The air bled from the engines not only keeps the interior warm but also forces the oxygen naturally present into the lungs of all on board. If the cabin pressurisation fails, a warning horn sounds and emergency oxygen masks drop down. Up on the flightdeck, pilots are trained to don their oxygen masks and then descend the aircraft rapidly to get it below 10,000ft where everyone can breathe naturally. If cabin pressurisation fails and oxygen masks are not put on, all on board will sink into a coma and die within minutes.

The urgency of such a situation is illustrated by the advice in the Learjet 35 Abnormal Procedures Checklist:

'If cabin pressurization cannot be maintained, execute EMERGENCY DESCENT as follows:
a. Oxygen Masks — Don. Select 100% oxygen.
b. Thrust Levers — Idle.
c. Autopilot — Disengage.
d. Spoiler Switch — EXTEND.
e. LANDING GEAR Switch — DOWN below [maximum operating limit speed] or [maximum landing gear extended speed] as appropriate for altitude.
f. Descend at [maximum operating limit speed] or [maximum gear extended speed] as appropriate for altitude. Descent from 45,000ft to 15,000ft requires approximately 2 minutes, 45 seconds.
g. Transponder — 7700 (Emergency).
h. Oxygen Mic Switches (Pilot and Co-pilot — On).
i. Notify ATC.
j. Check and Assist Passengers.'

On 25 October 1999, the flight crew's failure to respond to repeated ATC radio inquiries when the aircraft was climbing through 36,400ft was the first indication of an onboard problem. The Learjet should have turned left at Gainesville towards Dallas, but it made only a partial turn before heading in a straight line toward South Dakota. As the flight continued, it deviated from its assigned course and failed to level at its assigned altitude (FL390). Despite the instruction that flight crew must be able to don an oxygen mask within five seconds, there was no evidence that the flight crew attempted to intervene over the next four hours because the aircraft continued to fly off course. N47BA reached a maximum altitude of 48,900ft before finally descending to impact, even though the Learjet is only cleared to 45,000ft, which showed the extent to which there was no human activity on the flightdeck. From these events, NTSB investigators concluded that the flight crew became incapacitated at some point between 09.27:18 and 09.33:38.

The continuous sounding of the cabin altitude aural warning during the final 30 minutes of cruise flight (the only portion recorded by the CVR) indicated that the aircraft and its occupants experienced a loss of cabin pressurisation some time earlier in the flight. Set on autopilot, the Learjet flew 1,400 miles straight across half a dozen states before it ran out of fuel some four hours after it took off. Despite the severity of crash damage to the Learjet, there was no evidence suggesting any alternative reason for the incapacitation.

Two interrelated questions flowed from this. First, why did cabin pressure fail so insidiously such that the flight crew were unable to maintain consciousness? Second, why did the pilots not receive supplemental oxygen from the aircraft's emergency oxygen system to enable them to keep control and initiate descent to a safe altitude?

Post-accident examination showed the aircraft's pneumatic system to be intact and, therefore, normal system pressure was being supplied to the air conditioning system flow control valve. The F-16 pilots who observed the Learjet in flight reported that its windshield was obscured by condensation or frost, which was consistent with a loss of bleed air supply to the cabin. When bleed air is supplied to the cabin, the cockpit windshield receives a constant flow of warm air that prevents or removes condensation, regardless of the ambient temperature or pressure in the cabin. Thus, the windshield would remain relatively clear following depressurisation assuming continued bleed air supply to the cabin, but condensation could form and remain on the windshield following a depressurisation if there was a loss of bleed air inflow to the cabin. Therefore, it was very likely the Learjet did not have an inflow of bleed air to the cabin.

The Learjet has a flow control valve that regulates the rate of flow of the bleed air entering the cabin for heating and pressurisation. Investigators found this valve in the closed position in the aircraft wreckage, and they tried to find out whether this flow control valve may have malfunctioned and closed uncommanded. However, the condition of the wreckage prevented investigators from determining whether a mechanical malfunction had closed the flow control valve undemanded.

There was always the fear that the pilots had failed to select the CABIN AIR switch to NORM, which activated the air conditioning system (and thereby pressurised the aircraft), before take-off. Even though Item 19 of the Taxi and Before Take-off checklist specifies 'CABIN AIR SWITCH   NORM,' investigators observed that 'there is incentive to leave the pressurization system off during taxi and take-off in warm weather because inflow air can be hotter than cabin ambient air.' However, without the cabin air conditioning system, both pilots and passengers would have noted the effects of an unpressurised high cabin climb rate after take-off, not least because it would have caused discomfort. At about 10,000ft cabin altitude, the cabin altitude aural warning should have sounded, further alerting the flight crew to the lack of pressurisation. Although the pilots could have manually silenced the warning, they would have had to repeat this action every 60 seconds. At about 14,000ft cabin altitude, deployment of the passengers' oxygen masks would have provided an additional cue that the cabin was not properly pressurised. It was therefore unlikely that the flight crew would have continued to climb in the face of many clear signs that the Learjet was unpressurised.

With the CABIN AIR switch to OFF, the actions of the flight crew would have been impaired by hypoxia. Hypoxia occurs when there is insufficient oxygen in the blood and body tissue, leading to impaired vision and judgement, loss of motor control, drowsiness, slurred speech, memory degradation, difficulty in thinking, loss of consciousness and death. Yet the first officer showed no signs of hypoxia in her radio transmission at 09.27:18, when the aircraft was climbing through 23,200ft. Furthermore, the cabin altitude warning was not heard in the background of these radio transmissions. While it is possible that the frequency of the pilot's headset, the aircraft's radios or the ATC recording system may have masked the sound of the cabin altitude warning, the lack of such a sound suggests that the aircraft had not depressurised to a cabin altitude greater than 10,000ft by that time. Therefore, although the NTSB acknowledged that flight crew failure to activate the cabin air conditioning system before take-off was a valid safety concern for Learjet 35 operators, it considered this an unlikely occurrence on 25 October 1999.

Investigators considered the possibility that the flight crew inadvertently selected the CABIN AIR switch to OFF, thereby closing the flow control valve, during flight. N47BA was not equipped with automatic emergency pressurisation, so if the pilots had experienced a loss of cabin pressurisation from a cracked windscreen or whatever, they should have actioned the Learjet 35 Abnormal Procedures checklist. Step 4 of this checklist for a pressurisation loss at altitude instructs pilots to select the WSHLD (windshield) HEAT AUTO/MAN switch to AUTO, thus initiating emergency bleed air supply to the cabin. Step 5 in the same checklist instructs pilots to select the CABIN AIR switch to OFF, thereby closing the flow control valve.

Investigators considered the possibility that the flight crew might have experienced (or thought that they had experienced) a loss of cabin pressurisation. They then responded by trying to execute the abnormal procedure for a loss of pressurisation at altitude but omitted Step 4 (selecting the WSHLD HEAT AUTO/MAN switch to AUTO) before accomplishing Step 5 (selecting the CABIN AIR switch to OFF). Therefore, the closed position of the flow control valve could have been a consequence of the flight crew's attempt to address a pressurisation malfunction or failure (cause unknown), rather than its cause.

Overall, the NTSB deduced that an uncommanded closure of the flow control valve would have been sufficient to depressurise the aircraft. However, there was insufficient evidence to determine whether depressurisation was initiated by

a loss of bleed air inflow (caused by a malfunction of the flow control valve or by inappropriate or incomplete flight crew action), or by some other event. And that uncertain outcome is often the best that can come out of a heap of aircraft wreckage when there is no flight data recorder.

The second question was why, following the depressurisation, did the pilots not receive supplemental oxygen in sufficient time and/or adequate concentration to avoid hypoxia and incapacitation? A single oxygen bottle supplied the pilots and all passengers with emergency oxygen, and the wreckage indicated that the oxygen bottle pressure regulator/shut-off valve was open on the accident flight. Although one flight crew mask hose connector was found to be disconnected from its valve receptacle (the other connector was not recovered), damage to the recovered connector and both receptacles was consistent with both flight crew masks having been connected to the aircraft's oxygen supply lines at the time of impact. In addition, both flight crew mask microphones were found plugged in to their respective crew microphone jacks. Therefore, assuming the oxygen bottle contained an adequate supply of oxygen, supplemental oxygen should have been available to both pilots' oxygen masks.

At the crash site, the oxygen bottle was found to be empty. A Sunjet Aviation official told the Safety Board that Capt Kling had reported that the oxygen pressure gauge was in the green zone, indicating adequate pressure, during pre-flight checks on the day of the accident. However, even if the oxygen bottle had been full at the beginning of the fatal flight, the oxygen supply would have been completely depleted before impact because the regulator installed on one of the two flight crew masks would have automatically supplied 100% oxygen when the cabin altitude increased beyond 39,000ft. In such circumstances, a fully-charged oxygen

Above:
A typical Learjet
35A. *P. R. March*

99

bottle would have emptied in about eight minutes, so the Safety Board could not determine the quantity of oxygen that was on board N47BA. However, if the Learjet's oxygen bottle had been improperly filled with air, rather than oxygen, there would have been insufficient partial pressure of oxygen in the supplied mixture to avoid hypoxia at high cabin altitudes after a depressurisation. The Safety Board could not rule out the possibility that the oxygen bottle had been inadvertently charged with air instead of oxygen.

Another possible explanation for the failure of the pilots to receive emergency oxygen was that their ability to think and act decisively was impaired by hypoxia before they could don their oxygen masks. If there had been a breach in the fuselage (even a small one that could not be visually detected by the in-flight observers) or a seal failure, the cabin could have depressurised gradually, rapidly, or even explosively. Research showed that following rapid depressurisation at around 30,000ft it took as little as eight seconds without supplemental oxygen to significantly impair cognitive functioning and increase the amount of time required to complete complex tasks.

Investigations of other accidents in which flight crews attempted to diagnose a pressurisation problem or initiate emergency pressurisation instead of immediately donning oxygen masks following a cabin altitude alert have revealed that, even with a relatively gradual rate of depressurisation, pilots have rapidly lost cognitive or motor abilities to troubleshoot the problem effectively or don their masks shortly thereafter. On Payne Stewart's Learjet, the flight crew's failure to obtain supplemental oxygen in time to avoid incapacitation could be explained by a delay in donning oxygen masks of only a few seconds in the case of an explosive or rapid decompression, or a slightly longer delay in the case of a gradual decompression. None the less, the NTSB was unable to determine why the flight crew could not or did not receive supplemental oxygen in sufficient time and/or in adequate concentration to avoid hypoxia and incapacitation.

After the accident, President Clinton said, 'I am profoundly sorry for the loss of Payne Stewart, who has had such a remarkable career and impact on his sport and a remarkable resurgence in the last couple of years.' But five other precious people were also lost in that accident, and as airports and commercial airliners become more crowded, and travellers experience increased delays and cancelled flights, many more business travellers will choose to travel by corporate jets owned by their companies or on leased aircraft, or aircraft shared through fractional ownership.

The jet charter business uses the same high-performance jet aircraft and is growing just as rapidly as people want to fly between cities not served by direct flights, such as Brussels to Cincinnati, Nuremberg to Lyon. However, high performance places higher demands on aircrew and operators, which means that those flying on executive jets deserve to be as safe as those flying on major commercial airliners. But for everyone flying above 10,000ft, as most people do nowadays because that is where the air is calmer and jet engines perform best, it is worth remembering that none of us can survive up there for more than a minute or so without oxygen. Pay close attention to the flight attendant's emergency oxygen drill patter in future — it could save your life when others around you are losing a grip on theirs.

8

Keep Your Distance

The EP-3E Aries II (Airborne Reconnaissance Integrated Electronic System) is a conversion of the Lockheed Martin P-3C Orion maritime patrol aircraft. The US Navy maintains 11 of them, and Aries ushered in an all-new generation of Electronic Intelligence (ELINT) and Signal Intelligence (SIGINT) sensors linked together via a central processor.

EP-3E Aries No 156511 was one of a handful operated by VQ-1 'World Watchers' based at Whidbey Island Naval Station in Washington State. Every so often, one of them would deploy across the Pacific to Japan, and just before midnight on 31 March 2001, 156511 took off from Kadena Air Base in Okinawa as Mission PR32. The commander and pilot was Lt Shane J. Osborn, a hulking 26-year-old from Nebraska. On the flightdeck he was supported by a co-pilot, flight engineer and two navigators, all of whom were responsible for the safe operation of the aircraft.

The slow-moving EP-3E, powered by four turboprop engines, has a range of more than 3,000nm. Equipped with a distinctive saucer-shaped 'Big Look' antenna on its underbelly and loaded with electronic devices, it is more than capable of monitoring, intercepting and recording a wide variety of communications, including radio and radar transmissions, along the Chinese littoral and in the South China Sea. On 1 April, PR 32 carried no weapons.

Behind the flight crew stations were a mission crew of 19, many of whom would be long-serving intelligence and surveillance personnel. One of them was the in-flight maintenance technician. Then there was the Manual Electronic Intelligence (ELINT) operator, tasked with signals analysis and receiver control. He or she (there were three females on board) would take over mission co-ordination in the event of a main mission computer failure. Moving aft, the low-band signal collection station monitored early warning, height finding and meteorological radars. Next was the high-band analysis station, which monitored the airborne radars on interceptor fighters. Next to the high-band analysis position sat the operator who managed the 'Big Look' radar-optimised for maritime contacts and the aircraft's long-range electronic receivers, and who pulled together and evaluated newly acquired signals.

The ELINT supervisor sat next to the 'Big Look' position and supervised the Electronic Sensors as a whole. Next came the Electronic Warfare (EW) co-ordinator, who acted as master of ceremonies responsible for integrating acquired ELINT and Communications Intelligence (COMINT) data and disseminating it to off-board users. Then there was the COMINT special task supervisor/collection director who managed the aircraft's COMINT section. Located to starboard in the rear section of the main cabin, the COMINT console accommodated five operators whose tasks included COMINT collection, analysis and interpretation. The sixth COMINT position, located to port at the rear of the main cabin, was unoccupied on 1 April.

Late into the mission, the EP-3E had reached the airspace southeast of China's Hainan Island. The waters around Hainan are among the most strategically sensitive in the world, with various nations besides China claiming sovereignty to disputed areas such as the Spratly Islands between the Philippines and Vietnam. The main task of any EP-3E Aries is to provide the local US Navy Fleet Commander with up-to-date tactical information on the sort of threats he would face if his naval forces were drawn into conflict. Military intelligence is all about assessing an opponent's capabilities and intentions, and, while much information on fixed installations or in-use radio frequencies can be obtained from satellites and ground listening posts, a wily group of experienced watchers and listeners on board an aircraft is still the best way of assessing whether or not the opposition is using its equipment and facilities professionally or amateurishly, using its initiative or operating by rote.

An EP-3E operating over the South China Sea can be likened to a sharp stick that is pushed into the dragon's cave to see how it reacts. Not a few of PR 32's mission crew spoke Chinese and would know inside out the Chinese way of doing military business. Whatever they were tasked with that April Fool's Day, they got a reaction. Out of Lingshui airfield on Hainan Island hurtled two Jianjiji-8D ('Fighter aircraft 8D') 'Finback' all-weather interceptors belonging to the 9th Division, 25th Regiment of the People's Naval Air Force. The J-8D is an indigenously updated version of an old Soviet heavy fighter design, but it is no slouch and it packs a reasonable punch.

The two Chinese Navy fighters tracked and monitored the EP-3E. The lumbering EP-3E was heading in an easterly direction at around three miles a minute, but at this speed the Chinese fighters would have been close to the stall so one stood off while the other made a couple of passes. Lt Patrick C. Honeck was watching out of an EP-3E window at 09.07hrs [01.07 GMT]. 'After his first two runs at us, it got kind of surreal, like slow motion,' he said. The J-8 pilot, 33-year-old Lt Cdr Wang Wei, saluted the American crew on his first pass, and 'mouthed something to us' on the second. On the third approach, as they were 104km southeast of Hainan Island, the Chinese fighter collided with the American aircraft. The surveillance aircraft began falling — nose down — from an altitude of 22,500ft, while the 'Finback' appeared to break in half before crashing into the sea.

The collision ripped off the nose cone and took a chunk out of the port outer propeller. Consequently, the EP-3E snap-rolled left to an angle of bank of about 130°, which was getting near to inverted. The first thing Lt Shane Osborn thought was, 'This guy just killed us.' He remembered looking up, seeing water, 'and then I also saw another plane smoking toward the earth with flames coming out of it'. While Osborn tried to regain control of the aircraft, crewmembers behind him were terrified. 'The first thing I thought of was, "Oh, my God",' said Aviation Machinist's

Mate Second Class Wendy S. Westbrook. 'All I could see was blue water.' 'I was hoping and praying he was going to get us out of this,' Lt Vignery said. 'I didn't think we were going to make it. I had already accepted Jesus Christ as my Saviour, but I said another prayer at that time in case I didn't get it right the first time.'

156511 plunged 7,500ft (2,300m) in seconds. It was vibrating, the altimeter was out and warning lights were flashing. There are few windows on the EP-3E because most mission crewmembers need darkness as they huddle over their screens and oscilloscopes, so few down the back would have known that the J-8 was alongside until their world turned upside down. As the seconds passed, Lt Osborn gave the order to prepare to bail out. But as the controls responded and he was gradually able to pull the aircraft out of its dive, he told the crew to prepare for ditching instead.

It took the burly young Lt Osborn five minutes to regain control of his aircraft. The surveillance mission's senior chief petty officer, Nicholas Mellos, described that frantic time as 'mayhem'. Conditions made it difficult for crewmembers to concentrate, to breathe and to communicate. 'You had to yell to talk to the person next to you,' said Lt Honeck.

Lt Osborn regained enough control to enable Lt Honeck to move back to study maps to see where PR 32 might land. They quickly ruled out a run for Okinawa, which was four and a half hours away, or the Philippines, which was farther, because of strain on the wounded engine and fears that the damaged propeller might fly off and pierce the fuselage.

'I knew we were near Hainan Island,' said Lt Honeck, so the flight crew chose that as their safest destination, even though they did not have Chinese permission to land. Lt Osborn recalled that his crew made at least 15 distress calls, none of which was answered either by

Above:
An EP-3E Aries II, showing the underbelly 'Big Look' and other antennae that make it such a formidable electronic intelligence gatherer.
The Aviation Picture Library

Chinese ATC or by the surviving J-8 pilot. The noise inside the EP-3E was so loud that it may have prevented the Chinese authorities from hearing the Mayday calls, but the crew also dialled up the emergency 'squawk' on its transponder, which would have flagged up the distress state on any Chinese air traffic radar screens.

The EP-3E had lost its nose cone and, with it, the navigational radar and many other flight instruments. It was now clear that the errant cone had also disabled the starboard inner (No 3) engine, and an antenna had become wrapped around the tailplane. Osborn had no idea of either airspeed or altitude as the rush of air against the flattened nose buffeted his aircraft. With two suspect engines, Lt Osborn still doubted that he would be able to get the aircraft down safely. Up to the end, the mission crew feared that they were going into the water because no one knew the scale of the damage to the aircraft or its flying systems.

With the aircraft still well out from shore, Lt Osborn called for 'the emergency destruct plan'. The checklist read:

'In the event of emergency landing at a hostile airfield
1. Fill all hot water makers
2. Make and ensure maximum coffee yield
3. Water MUST be boiling
4. Pour all coffee over motherboards and into disk drives
5. Connect all computer busbars to power surge facility
6. Ignore all commands to evacuate until items 1-5 complete on all equipment
7. Light small fire on landing
8. On departure, fire Verey into cabin'

It is likely that the EP-3E was expected at the Chinese airfield. The pilot of the second J-8D tracking jet, after seeing the collision, had landed at the same field 10 minutes earlier. The runway appeared to have been cleared for the damaged aircraft's arrival, and as Lt Osborn signalled his intent to land with a 270° 'clearing turn' above the control tower, the electronic specialists in the belly of the aircraft were destroying as much sensitive data and equipment as possible. The Aries II made a hair-raising, flapless landing at Lingshui Airport, Hainan Island, at 09.33hrs. The Chinese authorities said that although the EP-3E entered Chinese airspace without approval, they allowed it to land in a spirit of humanitarianism.

After landing, 156511 was immediately surrounded by heavily armed Chinese troops while the US crew spent 15 minutes smashing and gutting. Soldiers peering through the windows yelled through bullhorns and pointed their weapons to make it very clear that the crew must stop. The US crew sent a final message home that they were 'closing up' and disembarking, whereupon they opened the door and were taken into custody. No shots were fired at any point. The crew had no alternative open to them. There was nowhere to go.

As aircraft commander, Lt Osborn wanted to be the first to walk off. A small group of armed Chinese military, including an interpreter, approached. 'He told us not to move and don't do anything. I asked if I could use a phone to call the US ambassador (in Beijing) to let him know we were safe on deck, but he said they had already taken care of that. Then they told us to get off the plane, and they were pretty adamant about it. We dropped a ladder, and I got off first.'

*   *   *   *   *

On the ground, the Chinese offered the crew water and cigarettes and 'told us not to worry'. The Americans were then herded into vans and taken to what they perceived as officers' quarters on the Lingshui military base. 'Their best barracks,' Lt Osborn said, 'but by American standards, they were poor. Lots and lots of bugs and mosquitoes. But it was liveable.' The crew were provided with basics like toothpaste and electric shavers, but no razors. Brig Gen Neal Sealock, an American military attaché from Beijing, got them clean socks and underwear. Sealock assured the crew that efforts by the Bush administration to secure their release were being conducted at the 'highest level' and that their families had been notified.

The crew remained at Lingshui for two nights before the Chinese moved them to a nearby base lodge in Haikou, a modest building where the Americans were placed into rooms on two floors, which the crewmembers assumed to be bugged. Lt Osborn was the only one to have a room to himself, which was not necessarily a luxury. Aside from meals, crewmembers were segregated from those occupying other rooms. The Chinese were polite and respectful but the Americans suffered from a lack of sleep and unpleasant interrogations, which lasted four or five hours the first night and later occurred with wake-up calls at all hours. 'I tried to steal some sleep when I could,' said Lt Osborn.

The rooms had television sets without cable and phones that did not work. Meals varied. 'It was Chinese food, but definitely not Americanised,' said Lt Vignery, adding that the Chinese served them fish heads 'until they realised we weren't into fish heads'. Some of the guards were friendly enough to provide decks of cards. Later, the guards brought them copies of a newspaper in English, from which the crew learned nothing of their predicament. But light reading was not enough to break the tedium, so Lt Honeck and Lt Vignery began writing humorous skits to perform in the hallways when the group was taken to their meals. 'They got quite a few laughs,' he said. 'We did a *People's Court* spoof, news like on *Saturday Night Live* and one of the cable show, *The Crocodile Hunter.*'

The crew eventually developed an amiable relationship with their guards. One wanted to know the lyrics to an American song he had heard, *Hotel California*, by the Eagles. It crossed the mind of at least one crewmember that two decades earlier, Americans were held hostage by their Iranian captors for 444 days before their release. 'But there was no wavering. I was confident we were going to get home.'

This stand-off phase was underlined by increasingly strident Chinese statements that 'the Chinese aircraft were flying normally when the US aircraft suddenly turned toward them and its nose and left wing bumped against a Chinese aircraft, causing the Chinese aircraft to crash. The Chinese side is searching the whereabouts for the pilot. We are much concerned about the condition of the pilot.' Tension rose when the Communist Party newspaper came out with a legal scholar's opinion that China was within its rights to confiscate the plane and charge the pilot. 'No one thinks they will actually do this,' confided a diplomat, 'but the Chinese are being very deliberate here in laying out their options.'

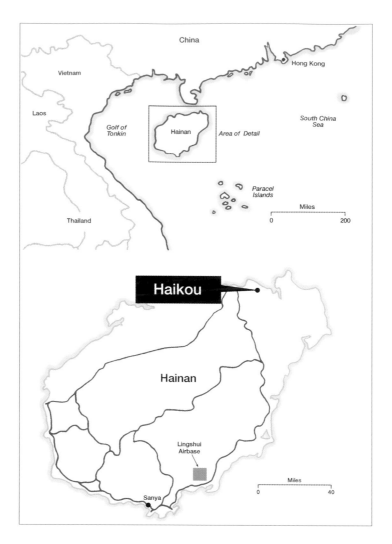

President Bush first heard about the incident from Condoleezza Rice, his national security adviser, who was a guest at the Camp David presidential weekend retreat in Maryland. The President's first decision was that, unlike his father, he would not 'reach out' to the Chinese. He did not use a hotline to contact President Jiang Zemin. He would be kept at arm's length from the negotiations.

In China the hawks were in the ascendant. After NATO bombers accidentally bombed the Chinese Embassy in Belgrade in 1999, killing three, many Chinese believed that the act was deliberate and complained that President Jiang had not stood up to American aggression. Lt Gen Xiong Guangkai, an army intelligence staff officer, was put in charge of a taskforce handling the crisis. The Americans knew Xiong's volatile mood and regarded him as dangerous. On one occasion he had serenaded a visiting American admiral with a rendition of *Edelweiss*. On another, he had threatened to drop a nuclear bomb on Los Angeles.

Chinese President Jiang Zemin, deeply concerned about missing fighter pilot Wang Wei's safety, gave repeated instructions to search for and rescue him at any cost. Over the days following the collision, the Chinese sent 48 aircraft and 29 ships to search for the missing Wang. The Americans' offer to assist in the search and rescue operation was not taken up.

The wild-card in the stand-off phase was how the Chinese authorities would deal with Wang's body if they recovered it. That had the potential to ratchet up the temperature, especially as the Chinese press was insistent that the EP-3E triggered the collision by moving 'suddenly'. It was only when all hope of recovering the Chinese fighter pilot's body was extinguished that the Chinese seemed to relent on holding the US crew.

After days of tense activity, the US Ambassador to China, Joseph W. Prueher, delivered the latest draft of a letter saying that Washington was 'very sorry' that a Chinese pilot had been killed in the collision, and it was also 'very sorry' for the unauthorised entry into Chinese airspace. That did the trick and the EP-3E crew were released on 12 April.

The detainees' festive homecoming was in contrast to two austere days of intense, no-nonsense debriefings in Honolulu, lasting up to 14 hours each day, as intelligence officials sought to learn more about the cause of the mid-air collision. Lt Osborn and the 23 other crewmembers of the downed EP-3E Aries II returned to a hero's welcome at their home base at Whidbey Island, where they were greeted by their anxious families and a cheering crowd of over 10,000 people. A few days later, a decree signed by the Chinese president posthumously named Lt Cdr Wang Wei a hero, 'a Guardian of Territorial Airspace and Waters'.

*   *   *   *   *

What caused the diplomatic incident in the first place? Both China and the US agreed that the collision occurred in international airspace about 105km southeast of Hainan Island, but that is all they did agree on.

The Chinese position was that at 08.36hrs Beijing time, the EP-3E approached Chinese airspace off the city of Sanya to conduct surveillance. The Chinese Navy sent two J-8 fighters to follow and monitor. At 09.07 the J-8s were flying normally on a heading of 110°. The US aircraft was flying to the right of and parallel with the Chinese fighters, which were between the EP-3E and the Chinese coast. 'The US aircraft suddenly veered at a wide angle towards the Chinese. The US aircraft's nose and left wing rammed the tail of one of the Chinese fighters, causing it to lose control and plunge into the sea. The pilot Wang Wei parachuted from his stricken plane, while the other Chinese plane returned safely and landed at 09.23.'

In response to the Chinese assertion that his aircraft made a sharp left turn into the Chinese jet, Lt Osborn said categorically that the turn happened only *after* his aircraft went out of control once his No 1 engine's propeller 'was impacted in the nose' of the Chinese fighter. The American position was that the EP-3E was

Left:
Map of the South China Seas, showing the area where the collision between the EP-3E and the J-8 took place and Lingshui airfield on Hainan Island.

lumbering along, and about to return to Japan, when the fighter closed to 'within 3-5ft' twice and finally struck the US aircraft while it was flying 'straight, steady' on autopilot away from Hainan Island.

The US claimed that the J-8D 'Finbacks' flew a straight course heading 110° in an effort to force the EP-3E away from Chinese airspace. According to Pentagon officials, Lt Cdr Wang Wei had flown dangerously close to American aircraft before and was known as a hazard. 'We have photos of pilots' faces. It was not the first time this individual had been that close to an aircraft. The number of intercepts and their aggressiveness has increased in recent weeks.' Such allegations of hot-dogging and harassing offended Chinese state media, which lauded the pilot as a hero martyred by reckless American actions.

International rules of the air are quite specific that a faster aircraft must give way to a slower, and a nimble fighter jet is much more manœuvrable than a slow-moving EP-3E the size of a Boeing 737. It is also the responsibility of any overtaking aircraft to keep a safe separation from the aircraft being overtaken. Under the international system of aerial cat and mouse that has evolved since 1945, shadowing aircraft are usually careful to stay at least 90m from the object of their curiosity and in full view of it.

If, as was claimed, Chinese pilots would regularly slip beneath US aircraft and then shoot up in front of them, either to put the EP-3E crew off their stride or show off Chinese naval aviation prowess, this was a very risky ploy indeed. Given the very controlled nature of Red Chinese military operations, where obedience is a virtue and individual flair and initiative is not always career-enhancing, it is hard to believe that the J-8 pilots were doing anything that had not been approved by higher authority. Although Chinese pilots had been coming to within 15m of US aircraft, such proximity should not have been life-threatening given that the EP-3E crew knew that the J-8s were there, and fighter pilots are trained to keep close formation. It is also standard practice for interceptor pilots to get close to take photographs of the opposition. The Chinese already knew much about the EP-3E's capability from analysing the external aerials, radomes, fairings and bumps on the wings and fuselage. Technical analysts can tell a lot about what is inside a pod from its dimensions, and Lt Cdr Wang may have been trying to get close to photograph some external feature when he overcooked it. On the other hand, given his history of getting close, he may have been trying to express Chinese dissatisfaction with the US surveillance presence just off their coast. To make his point, Wang may have roared up vertically in front of the EP-3E from underneath. If he was trying to shake the crew about, he succeeded beyond all expectations.

From the evidence it appears that the EP-3E was doing 180kt at an altitude of 22,000ft. Such an airspeed was close to the J-8's stalling speed, making it much less manœuvrable than normal. On the third pass, Lt Cdr Wang apparently miscalculated. The J-8's right wing came up, hitting the EP-3E's No 1 propeller. The tailfin of the F-8 then drove the EP-3E's port aileron fully up, causing the large aircraft to snap-roll near inverted at three to four times its maximum roll rate.

The fighter's nose impacted with EP-3E's radome, and this relatively glancing blow sprayed debris into the EP-3E's first and third propellers. The J-8 then broke apart. At this stage, the EP-3E's No 1 engine was flaming out, the radome had exploded and the aircraft had depressurised. All airspeed and altimeter information had been lost due to damaged or lost probes, and 156511 was vibrating violently due to the two damaged props and tailplane. The aircraft's high-frequency radio antenna had separated and wrapped around the elevator trim.

Given the extent of the damage, it took maximum effort from both pilots to bring the EP-3E under control. It had rolled to a 130° angle of bank with 30° nose down, and it was finally recovering at an estimated 15,000ft but still with a 3,000ft/min rate of descent. Even then, it needed 'cherry lights' (maximum power, or 'red-lining', on the three remaining engines) plus full right aileron initially to hold the wings level. Even after the aircraft's descent was finally arrested around 8,000ft, there were still serious doubts over the degree to which Lt Osborn could control the aircraft. The flight crew's greatest concern was separation of the port outer propeller due to high vibration, despite attempts to feather it. Given that the nearest friendly airfields were over 600nm away, Lt Osborn was right to opt for an emergency landing at Lingshui airbase. The ditching option, given the level of damage the aircraft had sustained and the tenuous degree of control maintained, would almost certainly have led to a number of the 24 crewmembers losing their lives. Being careful not to overfly land until he had Lingshui airfield in sight, Lt Osborn then overflew the runway at a perpendicular angle to check it was free of any obstacles and to make his intentions clear. He then turned the aircraft through 270° and made a 170kt, no flap, high weight (49,000kg), no trim, no airspeed indicator-working landing with a damaged left aileron, damaged elevator, high drag-inducing unfeathered No 1 propeller and full right aileron.

Lt Osborn did an outstanding piece of work in getting his crippled aircraft and crew down safely at a strange airfield after a horrendous experience. No surprise, then, that far from berating Lt Osborn and his crew for leaving a valuable surveillance asset in Chinese hands, his superior officers praised their airmanship, teamwork and conduct upon their safe return. 'Thank God for the training we do every day,' said a crewmember on their return to Whidbey Island. 'Without it, we'd be having a different press conference.'

After Lt Cdr Wang's J-8 broke in two, the Americans saw his seat eject and parachute open but believe he was killed instantly. The collision caused the EP-3E to swerve hugely. The second Chinese fighter pilot, in the split-second confusion, would have seen the EP-3E banking and his comrade's jet breaking up simultaneously, and assumed the surveillance aircraft caused the crash.

The Hainan incident has to be set in context. Russian air defence units monitor foreign reconnaissance aircraft operating close to the Russian Far East border every week, and nearly 1,000 such flights were detected and tracked in 2000. Up to 60 surveillance flights take place in the Far East every week, with at least China, Russia, Japan, Australia, Indonesia, Vietnam, Singapore and both Koreas conducting reconnaissance patrols similar to those flown by Lt Osborn's crew in Asia.

Questions need to be asked about whether these intelligence-gathering flights, which are provocative by their very nature, are worth the grief in this post-Cold War age. As a minimum, the US and China should now work out ways of avoiding such aerial mishaps in future, as they did in 1998 to avoid such incidents at sea. Turning to specifics, the collision on 1 April 2001 had all the appearances of having been a tragic accident rather than anything more malevolent. The flight safety lesson is that when military aircraft are shadowing each other, they must keep a sensible distance to prevent a repeat performance. If Lt Cdr Wang was trying to impress or intimidate the American crew, or both, he paid the ultimate price.

It is not clear precisely how much sensitive equipment and material was destroyed before PR 32 landed. The EP-3E crew had some 26 minutes between the accidental collision and touchdown on Hainan. They would have been trained in the rapid disposal of sensitive material and components in the event of an emergency landing away from main base, but all their training would have presupposed a degree of control and calm aboard the aircraft. There would have been little of that after the J-8D hit. No crewmember was badly hurt but it would be surprising if some of them had not been thrown about in the rapid descent from 22,000ft, which would not have engendered peace of mind and clear thinking.

The US Navy would have been told of the accident by radio straight away, so any code-books in the aircraft would have been worthless because the Pentagon would have cancelled them on the spot. There was also at least one shredder on board, which avoided any frantic chewing. Moreover, the EP-3E is fitted with a datalink so that intelligence can be passed securely and speedily to whomsoever wanted it. The best stuff would have been transmitted automatically long before 156511 landed.

The day after the mid-air collision, the People's Liberation Army Air Force sent a cargo aircraft loaded with men and technical equipment from Beijing to Lingshui airfield. They would have gone over 156511 with a fine toothcomb, examining the black boxes, aerials, sensors, integrating computers and datalinks to check that Chinese military intelligence was up with the US state of the art. Given that a Playstation 2 has more computer power than most of the kit on board PR 32, the biggest coup for Chinese intelligence was to piece together optimum US intelligence gathering techniques and flight procedures.

Some have likened the EP-3E to a vacuum cleaner. That is wrong. A vacuum cleaner is indiscriminate, 'hoovering up' both dust and diamonds with equal enthusiasm. Each EP-3E mission is very precisely targeted against whatever is causing US agencies most concern at the time. The Chinese would want to know on what the Americans were focusing on 1 April 2001, because that would reveal the gaps in US knowledge. The Chinese got to know everything there was to know about the inside of an Aries II before all $80m's worth of 156511 was returned in pieces to its rightful owner, courtesy of a giant Antonov heavy lift transport aircraft.

*   *   *   *   *

Less than a week before the Hainan incident, two USAF F-15Cs fighters took off around 12.30hrs from their base at Lakenheath in Suffolk, England, and headed for the Scottish Highlands. ATC lost radio contact with them just under an hour later.

Down on the ground a man and his wife saw the F-15C Eagles flying at less than 1,000ft. 'This jet came straight at us. I said to my wife, "this is extraordinary, it looks totally out of control," whereupon it veered way up into the sky, turned left, and went north toward Ben Macdhui into a heavy snowstorm. A second one came following it looking as if it was under control.'

The two fighters had touched briefly in mid-air and crashed into the mountains, even though their pilots were on the same squadron and had briefed beforehand on what each of them should be doing. Whether you fly close to friend or possible foe, it pays to keep your distance.

9

Gotchas

On Thursday 5 April 2001, a Boeing 737-3Q8 was parked on Stand D58R at London Heathrow after a flight from Brussels. Before the cabin doors could be opened, the aircraft was felt to shake slightly. The commander disembarked, suspecting that something had driven into his airliner. On walking around the right-hand wing he found that a ground services van had reversed into the 737's trailing edge. People in the vicinity told him that the van driver had reversed without guidance. It was 11.00hrs in broad daylight.

Which only goes to show that human beings are capable of making mistakes. Since 1955, military and civilian flight safety staffs have worked hard to reduce the number of flying accidents attributable to mechanical failures, but notwithstanding all the advances in education techniques and behavioural science, the percentage of accidents attributable to human failings has stayed pretty constant. Try as they might, flight safety experts have never been able to overcome the basic design failing in human beings that, no matter how intellectually gifted or adept they may be, they can do silly things under pressure. Worse than that, they may create that pressure for themselves. Either way, there are a whole host of 'gotchas' out there in aviation just waiting to trap the unwary.

Back on 18 December 1997, a VC-10 C1K airliner-cum-tanker was minding its own business on the pan at RAF Brize Norton. XR806 was scheduled to be defuelled before going into the hangar for servicing, a procedure that had been carried out by skilled tradesmen many times before. Only on this occasion — perhaps because those involved were looking forward to some pre-Christmas cheer — the VC-10's main fuel tanks were emptied before the tail. All the school physics lessons about force x distance and centres of gravity came into their own and XR806 tipped gently back on its tail. Two people stuck in the cargo hold had to be rescued by the RAF Fire Service. Worse than that, the VC-10 suffered such damage around the auxiliary power unit and to the integrity of the pressure bulkhead that it had to be written off. An expensive flying asset was broken up for spares in March 1999, and all because someone's brain was left in neutral.

Some incidents would be almost hilarious if they weren't so potentially life-threatening. In the 1980s, when Saddam Hussein was a friend of the West, a four-ship of Mirage F1 fighter-bombers night-stopped in Greece outbound from France to Iraq. A bored Greek conscript sentry took it into his head to swing on the pitot tube of one aircraft, which bent gently downwards. He couldn't bend it back, so went along the line swinging from all four, in the hope that if they all looked the same, no one would notice. Yet another example of how aircraft can survive many hundreds if not thousands of flying hours without an aircrew-induced scratch, only for a perfectly serviceable aircraft to come to grief at the hands of unthinking 'support' personnel!

But when aircrew do screw up, they usually do it in spades. Fifty-odd years ago, the leading British jet fighter was the Gloster Meteor. From 1946 onwards, 658 of the F.4 type entered RAF service, a variant that was also sold to the Argentine, Belgian, Dutch, Danish and Egyptian air forces.

The Meteor conversion unit was at No 203 Advanced Flying School at Driffield in the East Riding of Yorkshire. In the summer of 1951, 203 AFS was flying from the huge Carnaby relief landing ground to the west of Bridlington while Driffield's runway was being repaired. The weather had been poor in the morning, but the stratus cloud lifted after lunch and full flying was resumed. A formation took off, led by a very experienced instructor with two Dutch students as wingmen.

About 30 minutes later the cloud started to roll in again off the North Sea and a general recall was broadcast on Driffield Approach. The only approach aids available were a manual VHF/DF feeding into GCA. The manual homer required long transmissions (several seconds) to get a reliable bearing, so only about five or six aircraft could be handled at a time on a Controlled Descent. Meteors had an endurance of less than one hour, so Driffield approach was under pressure from solo aircraft declaring low fuel states. The formation leader therefore elected to do a free letdown with what help he could get from the Fighter Fixer Service (VHF/DF triangulation) and Carnaby Homer. He positioned his formation over the sea, well northeast of Carnaby (so he thought) and began a descent through cloud on a southerly heading. His wingmen, although new to the Meteor, were experienced fighter pilots and steady in formation. The leader expected to break cloud above 500ft over the sea, wait to obtain a true bearing from Carnaby greater than 090° and then turn right onto West to run in over Bridlington, where the coast was low-lying. Unfortunately the winds at altitude were probably weaker than forecast and he started his descent closer to the coast than he thought. The coastline north of Flamborough Head runs NW/SE and the cliffs rise to about 250ft above sea level. The leader did not gain visual contact with the sea until about 150ft in poor visibility, unaware that he was closing with the cliffs at an angle. He saw the cliffs through the murk, yelled 'Pull up!' on the radio, scraped over the top himself, but his wingmen powered into the cliffs. The marks are still there.

At least the Meteor pilots were talking the same language. On 25 May 2000 a Streamline Aviation Shorts 330 freighter bound for Luton entered Runway 27 at Charles de Gaulle airport, Paris, at the same time as an Air Liberté McDonnell Douglas MD-83, with a crew of seven and 155 Spanish football fans on board, was taking off on the same runway. The MD-83 was travelling at 170mph when its left wingtip ripped through the right side of the smaller 330's cockpit, killing the co-pilot Jon Andrew. The captain, Gary Grant, who was sitting beside him, was seriously injured. It was so close to being a major disaster.

Under guidelines set down by the International Civil Aviation Organisation, English is recommended as the common language of aviation. That said, countries are allowed to opt out and use their native language with their pilots if they consider it safe. France and Russia are among a handful of countries that take advantage of the lack of a binding regulation enforcing the use of English.

The radio exchanges between ATC and MD-83 and Shorts 330 pilots on 25 May highlight the problems that this fudge can generate:

Controller: '*Liberté quatre-vingt-huit zéro sept, autorisé décollage vingt-sept, deux cent degrés, dix à quinze noeuds.*' A few seconds later, the controller switches over to English and transmits. 'Streamline 200, line up runway two seven and wait, number two.'
French captain: '*Tu es prêt* . . . are you ready?'
French co-pilot: 'I'm ready.'
British co-pilot (responding to controller): 'OK.'
British captain: 'Where's the number one? Is he the number one?'
(30 seconds later) 'Can you see anything down there?'
British co-pilot: 'No I can't . . . Unless there's one coming out in front.'
British captain: 'How about now?'
British co-pilot: 'S**t!'
French captain: 'Oh *putain!*'

It is clear from these CVR tapes that the British crew were cleared to line up and wait as 'number two'. The controller thought the aircraft were on the same entry point, but they were on different slipways and neither could see the other until seconds before the collision. The British captain was clearly confused because he was

Above:
A tandem-seat Meteor T.7 similar to those flown from the Advanced Flying School at Driffield.
Author's collection

twice heard asking his co-pilot where the one ahead was. If he had understood French, he would probably have realised that the Air Liberté MD-83 had just been cleared to take off from the same runway.

The Bureau Enquêtes-Accidents said that the British crew's inability to understand the instruction in French helped to cause the accident. Consequently, the BEA recommended that the French civil aviation authority, the DGAC, should consider using only English for air traffic control at its major airports.

Only a month before the accident, Air France had suspended plans to make English obligatory at its Paris hub after fierce opposition from pilots and politicians. It is all too easy to blame this accident on the French practice of using both languages in their airspace, but that is something of a cop out. Non-French speaking pilots operating in France have to be extra alert to what is going on around them, and there are plenty of airports around the world where the 'English' spoken is far from easily intelligible to native English speakers.

With all due respect to the Bureau Enquêtes-Accidents, the British crew's inability to understand instructions in French did not help to cause the accident. Knowledge of French might well have prevented disaster, but the accident itself was caused by both ATC and pilots losing the plot. If the ATC controller had kept a grip of where the aircraft he was controlling were, and if the Shorts 330 pilots had refused to enter an active runway until they knew exactly where the 'number one' ahead was, this accident would never have happened. The ability to retain their situational awareness, irrespective of the chatter going on around them, is what differentiates survivors from the rest.

'Gotchas' tend to creep up on a pilot. On Friday 3 November 2000, G-BYDN, a Fokker F100, was on the fifth leg of the day flying between Newcastle and Paris, Charles de Gaulle. The First Officer, who was undergoing line training, was in control of the aircraft from Newcastle to Paris, and there were no visible signs of icing.

The skies were clear during the initial descent into Charles de Gaulle. On the descent, Autopilot 2 (AP2) was engaged and the F100 was navigated by the Flight Management System (FMS).

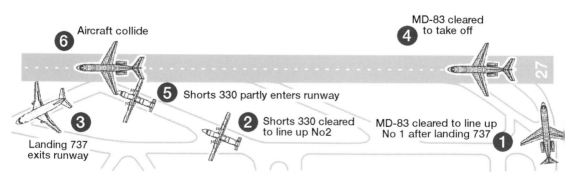

6 Aircraft collide

4 MD-83 cleared to take off

5 Shorts 330 partly enters runway

3 Landing 737 exits runway

2 Shorts 330 cleared to line up No2

MD-83 cleared to line up No 1 after landing 737

1

G-BYDN was levelled at FL110 and turned on to the 100° VOR radial. At this point, the F100 began to oscillate gently in pitch which increased as the aircraft gradually descended. The commander told the FO to de-select the autopilot, and, taking manual control, the first officer found he had approximately only 2cm of fore and aft control column movement, which rapidly reduced to no movement at all in pitch. The commander took control and confirmed that the column had jammed: only by pulling back firmly was he able to raise the nose of the aircraft sufficiently to climb back through FL 110. ATC queried the level variations and was told of the control difficulties. Both pilots had to push on the control column and, given the deteriorating situation, the commander transmitted a Mayday call.

Around this time, Autopilot 1 (AP1) was selected though neither pilot remembered making the selection. ATC acknowledged the Mayday and told the crew to turn right and descend to 3,000ft.

The Fokker began gently to pitch up and so the pilots pushed the control column forward. With speed reducing, it took both pilots to hold the control column in the almost fully forward position. With commendable presence of mind, the 49-year-old commander told the cabin crew to move all 71 passengers to fill up the front seats to help pitch the

Left:
Chart illustrating the fatal confusion at Paris-Charles de Gaulle on 25 May 2000.

Above:
An MD-83 in Air Liberté livery.
The Aviation Picture Library

115

aircraft down, and to prepare for an emergency landing. These rather disturbing instructions were carried out promptly and in an orderly manner.

With the control column nearly fully forward and the air speed at 232kt, the commander lowered the first stage of flap (8°). The pilots were now able to relax the forward pressure on the control column and the commander, having at various times tried to operate the electric trim with what appeared to be little effect, moved the manual trim wheel to try and trim the aircraft more nose down. The commander flew a radar vectored ILS approach to the nearest runway, 27 Right, and made a normal landing at 19.48hrs. The F100 was halted with reverse thrust. The control column electric trim switch was operated after landing, and it was found to be functioning normally.

Analysis of recorded flight data revealed two distinct periods of pitch instability during the incident — the first at the end of the descent to FL110, and the second during final descent from FL110. Starting with the first, as the F100 descended through 11,160ft at 19.35hrs, a target altitude of 11,000ft was captured and the autopilot's goal changed from 'vertical speed' to 'altitude hold'. In response, the autopilot applied a small amount of nose-up elevator and the aircraft pitch attitude increased smoothly from nose-down to nose-up. The autothrottle increased engine thrust to maintain 280kt.

As the F100 approached 11,001ft, a small amount of nose-down elevator was applied to arrest the pitch-up and level the aircraft. During the next 24 seconds the aircraft gently descended, reaching 10,973ft 16 seconds after the elevator ceased to move. The autopilot should have responded to this deviation by applying nose-up elevator but none was observed. However, the stabiliser, which was designed to trim out a constant elevator deflection, did begin to trim in the nose-up sense. This action increased pitch attitude to 1.25° nose-up and the aircraft began to climb back towards FL110. In an attempt to level at FL110, the autopilot should have responded with a nose-down elevator deflection but again none was observed. However, the position of the stabiliser began to respond in a nose-down sense, decreasing the pitch attitude of the aircraft in an attempt to level at 11,000ft.

Under the influence of stabiliser trim, the pitch of the aircraft continued to reduce through a level pitch attitude as the aircraft reached a maximum of 11,002ft and began to descend again. During the following two and a half seconds, the aircraft descended to 10,923ft and the stabiliser began to trim in the nose-up sense. At that point the elevator position changed rapidly from nose-down to nose-up, followed 1.25 seconds later by the F100 pitching nose-down. No other instances of anomalous elevator behaviour were observed during the remainder of the flight.

The pilots manually disconnected the autopilot at this stage, made a brief application of nose-down elevator and then maintained an average of 1° nose-up elevator as the aircraft pitched up. As pitch attitude increased through 2° nose-up and the elevator position remained relatively constant, the stabiliser trimmed nose-down. Pitch attitude reduced and the aircraft began to descend. A further input of nose-up elevator arrested the pitch-down at an attitude of 2° nose-down. During the next 30 seconds, a reducing amount of nose-up elevator and progressively more nose up-trim were applied as the aircraft climbed back towards 11,000ft. Throughout, the auto-throttle system strove to maintain a selected airspeed of 280kt. As the Fokker became more stable in pitch, the flight recorder showed that a reduced air speed of around 260kt was selected. The crew reported that considerable force was required to move the control column in pitch throughout this period.

In attempting to correct the unnerving oscillations in pitch, the pilots were under the impression that the elevator could not be moved following de-selection of AP2. Both pilots had to use extreme force on the controls in order to try and overcome the control restriction, but the commander was aware of other traffic 1,000ft below his aircraft and he put on his landing lights to make the F100 more conspicuous.

Following the incident, G-BYDN was examined closely and all the cables, motors, actuators, flight control computers and the flight mode selector panel were found to be blame-free. What had happened was that for 24 seconds in the descent down to FL110 when elevator movement was severely restricted, Autopilot 2 was controlling the aircraft's pitch attitude by the stabiliser alone. The elevator restriction was most probably due to ice accretion on the servo capstans and cables. The worst place for icing is between +3° and –5°C and although the crew saw no signs of icing on the outside of the F100, they were right in the middle of the crucial band as they approached FL110: the outside air temperature, which had been progressively increasing during the descent, reached 0°C at 11,200ft and +1° C at 11,001ft. The F100 had climbed through a layer of cloud on its departure from Newcastle, and the implication was that moisture picked up in the process had frozen during the descent on that wintry November evening. The symptoms were consistent with other reported cases, but because the evidence disappeared once the aircraft descended into warmer air, this could not be positively established.

Shortly after G-BYDN became stable at FL110 and the crew declared an emergency, Autopilot 1 was engaged. Pitch attitude was maintained by trim and elevator, but no significant variations were recorded during this phase of level flight, which lasted for approximately one minute.

The second instability phase started as the F100 left FL110. After the crew's Mayday call, ATC instructed them to turn right onto 120°M and descend to 3,000ft. At the start of the heading change from 100°M, thrust on both Rolls-Royce Tay turbofan engines was reduced and the airspeed started to fall from 260kt. Nine seconds later, still in a right turn and with airspeed reducing through 255kt, the autopilot operational mode altered to 'level change' and the speed mode changed to 'elevator controlled airspeed'. These mode changes were consistent with a descent to a lower level under autopilot control, although, believing the autopilot to be deselected and the aircraft under manual control, these selections were made using the flight director. A small amount of nose-down elevator and pitch trim was recorded as both engines started to reduce towards idle. As pitch attitude reduced and the aircraft began to descend, the crew again advised ATC that they had control difficulties and requested a priority landing.

Over the following 30 seconds, progressively more nose-up elevator and more nose-down pitch trim were recorded. In addition, more fly-left rudder (up to a maximum of 6.4) was applied as the aircraft tried to complete the right turn onto 120°M, and aileron deflection increased in the roll-right sense to compensate for the applied rudder. The crew again emphasised that they had a major control problem and requested radar vectors for a visual approach to any runway. ATC responded with a clearance to 27 Right.

At the end of the 30-second period, elevator position was recorded as being 8.7° nose-up, the stabiliser had stopped moving at a position 2.65° nose-down, airspeed had reduced to 250kt and the aircraft had descended 1,000ft to FL100. The response of the pitch trim system during this phase was consistent with an attempt by the autopilot to reduce altitude and achieve a target selected airspeed of 250kt whilst being opposed by a force on the control column in the aft direction. The behaviour of the aircraft in roll was consistent with an attempt by the autopilot to turn the aircraft to the right under aileron control whilst being opposed by a rudder deflection in the opposite direction.

What was happening was that the pilots were fighting the F100's automatics. G-BYDN was fitted with a Rockwell Collins duplex Automatic Flight Control and Augmentation System (AFCAS) that controlled the aircraft in pitch, roll, yaw, speed and thrust. The AFCAS consisted of three subsystems: the Automatic Flight Control System (AFCS), which provided the autopilot, flight director and altitude alert functions; the Auto-throttle System (ATS); and the Flight Augmentation System (FAS) which looked after the yaw damper and stabiliser trim.

The pilots' control columns were connected via cables to a pulley wheel and cable tension regulator assembly located in the bullet fairing on top of the stabiliser. An output rod from the pulley provided an input to a dual hydraulic actuator, either of which was capable of operating the elevator. Two autopilot servomotors (one for each autopilot) were located aft of the cable tension regulator, and drive cable capstans connected to the same pulley assembly as the pilots' control columns. The stabiliser was operated by two hydraulic actuators. Trim was usually controlled by the FAS whilst an autopilot was engaged. In manual flight, trim was controlled by switches on the control column yokes. The FAS disabled these when an autopilot was engaged, and, in the event of an FAS failure, a manual trim wheel on the pedestal did the business.

Everything should have got back to normal as the restrictive ice melted in the descent, but the 'gotcha' was that someone inadvertently selected Autopilot 1 some 78 seconds after AP2 was disengaged. This went unnoticed, probably because the pilots were preoccupied with arresting the aircraft's descent from FL110. The trim switches on the control columns would therefore have been isolated as the commander decided to reduce speed to 250kt. Neither pilot could recall selecting AP1, but from that selection until movement of the manual trim wheel, which automatically disconnected AP1, there was an ongoing struggle to control the aircraft's flight path.

According to the manufacturer, 'If the autopilot is "told" to maintain a pre-selected altitude it will go to any length to do just that. If the pilot overpowers the system using his elevator, the autopilot will "correct" this input by using the stabiliser . . . As the stabiliser has more authority than the elevator, the autopilot will eventually be able to climb to the pre-selected altitude again while the pilot is still pushing in order to descend.'

In other words, an engaged AFCS will always overcome any manual control inputs due to the greater control power of the stabiliser. The electric trim worked only with the AFCS deselected and, consequently, the commander's attempts to use the electric trim were destined to fail when AP1 was connected.

The upshot would have been humorous if it hadn't been so unsettling for the crew of five and the 71 passengers on board. Instead of descending in stately fashion, the F100 oscillated all over the place as the pilots and automatics were out of synch with each other. As the airspeed was reducing towards 245kt, the stabiliser started to trim nose-up and, as pitch attitude began to increase, less nose-up elevator was applied. Once airspeed had reduced to 245kt, stabiliser trimming ceased for eight seconds and the elevator position remained relatively constant at 3° nose-up. There then followed a period of continuous nose-up stabiliser trimming which lasted for half a minute, during which the stabiliser moved to -4.9° and the elevator moved in the opposite direction to 14° nose-down. Airspeed started to reduce towards 235kt as pitch attitude increased from 4° nose-down to 1.6° nose-up, and the aircraft descended through 8,700ft. As G-BYDN's heading approached 120°M, less left rudder was applied and roll-right aileron deflection was reduced. The pitch trim response during this phase was consistent with an attempt to target a selected airspeed of 230kt whilst being opposed by a force on the control column in the forward direction. Over the next minute, airspeed varied between 231kt and 236kt as stabiliser positions of between -3.9° and -5.9° (more nose-up), elevator positions of between 10° and 16° (nose-down) and pitch attitudes of 0.8° nose-up to 2.7° nose-down were recorded. During the last 30 seconds, the predominant trend was more nose-down elevator against more nose-up trim.

G-BYDN was now at 6,850ft and, as a further nose-down input was bringing the elevator position to 17° nose-down (close to maximum deflection), the first stage of flap was selected. During flap extension the stabiliser trimmed in the nose-down direction to -4.9° and pitch attitude reduced to 4.4° nose-down. Less nose-down elevator was applied and pitch attitude increased. The hitherto fluctuating airspeed reduced towards 220kt. Over the next 30 seconds, progressively less nose-down elevator was applied and the stabiliser moved in an increasingly more nose-down direction.

As the aircraft descended through 5,600ft, the stabiliser position moved in the nose-down sense. The movement was much quicker than that previously recorded, which was put down to a response to a crew trim input. Pitch attitude reduced by 2.5° to 4.5° nose-down before less nose-down elevator was applied, raising pitch attitude to an average of 3° nose-down. Whilst maintaining this pitch attitude, two further nose-down stabiliser movements were recorded. The second of these was coincident with the commander moving the manual trim wheel, which automatically disconnected AP1. From now, the pilots were back in control.

Over the next 30 seconds aircraft attitude remained relatively constant as the stabiliser was manually trimmed progressively more nose-down, and the elevator position moved progressively more nose-up. Airspeed increased to 230kt and after a further 10 seconds, during which the control surfaces remained relatively constant, the stabiliser was manually trimmed in one step to 0° (more nose-up), pitch attitude increased to level and less nose-up elevator was applied. The trim systems were re-engaged and both the stabiliser and elevator surfaces were brought to a mid-range position. This damping of the oscillations coincided with receipt of ATC radar vectors for a right turn to intercept the ILS for 27R.

The remaining five minutes of the flight were uneventful. Touchdown occurred at 19.48hrs with flap selected at an airspeed of 130kt. Reverse thrust was used to slow the aircraft. Fortunately, there were no injuries to anyone on board and no damage to the aircraft.

Why didn't the F100 commander, who had 6,900 flying hours of which 1,000 were on type, realise what was going on? It may have been that the pilots, who had flown the Newcastle-Paris route several times that day with no sign of any icing, never considered that icing would initiate their problems. It is unlikely that they were aware of the implications of ice accretion on the F100's servo capstans and cables.

Turning to the inadvertent Autopilot 1 selection, which was the cause of the post-icing difficulties, in front of each pilot were a Primary Flight Display and a Navigation Display screen, mounted one above the other. The upper screen displayed attitude, airspeed, altitude and flight mode information. The Flight Mode Annunciator (FMA) at the top of the Primary Flight Display screen gave five columns of information — thrust (THR), speed (SPD), flight path (PATH), lateral navigation (LAT) and autopilot, autothrust and flight director engaged status (STATUS). The lower Navigation Display screen provided horizontal situation information including heading and navigation data. Two Multi Function Display Unit screens were mounted on the lower part of the instrument panel between the two pilots. The left-hand screen displayed systems information, and the right-hand screen showed normal and emergency checklists.

AFCS selection was by pressing switches located on the glareshield above the instrument panel. The autopilot selection switches were marked AP1 and AP2. They had a glass insert which, when an AP was selected, illuminated a green light of adjustable intensity to show which AP was active. When auto-thrust was engaged on 3 November, the letters AT would have been shown in white below the AP information on the FMA STATUS column.

When the AFCS components were removed from G-BYDN and tested, no defect was found that would have caused AP1 to engage without normal selection by a crewmember. Following engagement of AP1, the aircraft would have been in navigation mode using the Flight Management System (FMS). Level change could be activated either by selecting a new level in the altitude window on the AFCS control panel and pressing the level change switch, or selecting the new altitude by pulling the rotary selector knob to activate the level change mode. With AP1 selected, the AFCS would have made the F100 follow the level and heading change commands. AP1 in white would have been shown in the right-hand column with AT in white below it.

So there were plenty of signs that AP1 was selected and trying to run the show. The 'gotcha' was that the flight crew thought they faced one serious control problem when in fact they were dealing with a transitory icing restriction followed by an inadvertent autopilot selection. Selection of AP1 gave the appearance of continuing control difficulties, which increased the pilots' concerns and workload at a critical phase of flight in a busy terminal area.

Although indications that AP1 was in control were all around them, the pilots' preoccupation with trying to control their aircraft meant that these were not noted. The comment by the FO that he could not see a green light in the AP1 select switch might have been due to its brightness having been turned down, but human factor studies have shown that, at times of heavy workload and in emergency situations, it is possible for pilots to be unaware of both visual and aural alerting devices. Under stress, tunnel vision takes over and cues outside the tunnel get ignored. After investigation of the G-BYDN incident, Fokker was asked to draw F100 operators' attention to the possibility of ice accretion on elevator servo capstans in cold humid conditions, and to introduce a revised capstan groove. But this was something of a cop out, especially as the Dutch believed that accretion of ice on the servo capstan could not be positively established. If G-BYDN had come to grief, it would have been because the flight crew got themselves into severe difficulties *after* the ice had melted.

It is arguable that work ought to be done into the potential safety implications of overpowering an autopilot. But of wider importance is the trend away from manual flying and for greater and greater reliance on automatics. All the glass panels and computer wizardry on modern flightdecks are very reliable and labour saving, but they do not absolve flight crews from *understanding* what is happening around them. An F100 crew should know that any attempt to try and fly the aircraft with the AFCS engaged will result ultimately in a victory for the AFCS. Realistic simulator training would teach that the Flight Director bars show any deviation from the commanded flight path caused by the pilots trying to counter the AFCS.

In bygone days, conversion-onto-type courses included getting to grips with how flight control systems and flight instruments worked, and their interrelationships. Now, if new pilots ask how something works, the response is likely to be, 'It works very well.' If airlines are cutting ground school training to the bone, they should remember that no matter how advanced aircraft and their systems become, proper flight safety is all about understanding what is going on inside the aircraft and the impact of its surrounding meteorological and air traffic environment.

The F100 commander was on the ball when it came to using talking ballast in the passenger compartment to bring the centre of gravity forward, and to put his landing lights on to warn any other aircraft to keep their distance. Moreover, the pilots of G-BYDN would have had their hands full with what they thought was a severe flying control problem, leaving neither spare arms nor opportunity to refresh themselves on the systems manuals. But in general, when a flight crew is under pressure, someone should keep a grip on the big picture. If at all possible, the commander should leave the flying to the first officer and stand back mentally from the minutiae to best assess what is going on, and the right course of action can then be taken.

The F100 pilots were caught out on 3 November by an unusual combination of circumstances. Most 'gotchas' result from human frailty, and to prove that there but for the grace of God go I, back in January 1971 I was co-pilot on the first Victor B.2 (SR) to be sent to Hong Kong. In those days when we could not overfly the Indian subcontinent, we could not make Masirah to Singapore in one

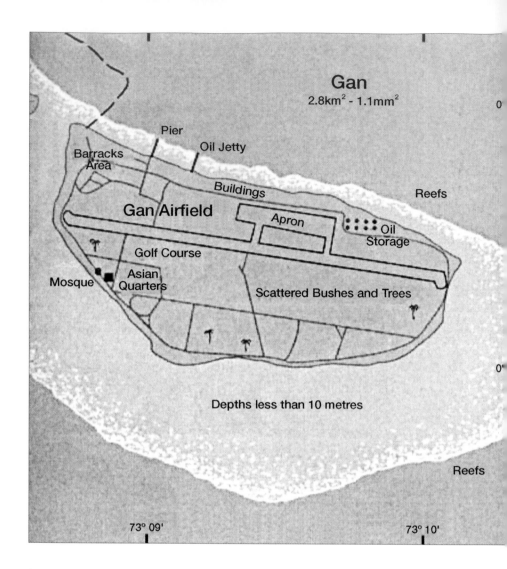

Gan
2.8km² - 1.1mm²

Pier

Oil Jetty

Barracks Area

Buildings

Reefs

Gan Airfield

Apron

Oil Storage

Golf Course

Mosque

Asian Quarters

Scattered Bushes and Trees

Depths less than 10 metres

Reefs

73° 09'

73° 10'

hop so we staged through Gan in the Maldives. This small coral island on the Equator was big enough for only a 7,500ft runway; there was no room for a taxiway, so landing aircraft had to roll to the end of the runway, turn around the dumb-bell and backtrack.

We had to arrive over Gan with enough fuel to stand off in the event of a tropical thunderstorm overhead, and to divert to Sri Lanka if the need arose. Having pitched up at Gan at close to maximum landing weight, we should have streamed the tail parachute on landing. In those days before thrust reversal was standard, the Victor carried a tail parachute to kill landing speed after

122

Above:
Gan airport, showing how little margin of error there was at the end of each runway before running into the Indian Ocean.

touchdown. The tail 'chute was very effective but it was a laborious pain to repack it after landing. Although the rules were very specific — streaming the chute was mandatory on runways less than 9,000ft long — if you landed at the right speed on the piano keys, experience had taught that you could get away without streaming in much less than 7,500ft and be off for a refreshing shower and glass of jungle juice pdq.

We crossed the piano keys a little hot, but we didn't stream because we were sure that the brakes would cope as always, as speed dropped below 100kt. Unfortunately, with half the runway gone and just 3,500ft to go before the Indian Ocean, it became clear that the hydraulics had chosen that day to fail. Only by streaming far too late and cornering the Victor round the dumb-bell on the nosewheel steering did we just avoid a career-finishing, unscheduled arrival in the Gan lagoon. Sod's Law says that a failure will happen on the day you try and bend the flying rules. The F100 incident showed that unsolicited 'gotchas' can creep up at the best of times — it is foolhardy to give them a helping hand.

10

Crowded Approach

At 07.11hrs on 16 September 2000, European BAC One-Eleven G-AYOP left Manchester International Airport for Bergamo in Italy. Two minutes 1 later ATC cleared the aircraft to line up on Runway 24R behind an inbound Airbus A320. Further out on the ILS approach was a Singapore Airlines Boeing 747-412 'Megatop'. The B747 crew, inbound from Amsterdam, told the tower controller that, 'We're heavy and we'll be using the whole runway.'

The A320 landed but was slow to turn off the runway. The controller told the crew to 'Expedite to the right and contact Ground (on frequency) 121.85.' .He then told the One Eleven crew to be 'Ready when cleared'. The crew of the A320 were slow to clear the runway so the controller transmitted 'Could you expedite the rapid exit'. The A320 pilots said that they would.

As the A320 took the second rapid exit turnoff (RET), Tower cleared the One Eleven, which by now was lined up, for take-off. The inbound 747 was then approximately one mile from touchdown and the controller asked the 747 commander if he could see the 'rolling' One Eleven ahead. The 747 commander said that he could. The controller noted that 'the One Eleven's acceleration appeared to be very slow whilst the approaching 747 appeared to be surprisingly fast so much so that it would reach the start of the runway whilst the One Eleven was only some way along it'. Nevertheless, the controller cleared the 747 to land 'after the departing' as he felt that was the safest option.

Turning his attention back to the airliner waiting to begin its take-off roll, the controller cleared the One Eleven to take off. As the One Eleven began its take-off roll and its twin Speys reached full chat, the 747 'Megatop' was only 0.5nm from the runway threshold. At 07.16hrs the 747 pilot decided that enough was enough and he decided to go-around. As the 'Megatop' crossed the runway threshold markings, its commander poured on the coals approximately 50ft above the ground, transmitting as he did so, 'On a go-round . . . and we'll start a right turn to miss the traffic.' The One Eleven, now some 400m ahead of the 747, had not yet reached rotate speed. The controller replied, 'Roger, that's a right turn.' As the 747 began to climb and turn slightly away from the departing One Eleven, which by now was airborne and climbing slowly, the 'Megatop' commander asked, 'Tell us what you advise, we cannot see that traffic.'

The controller replied that the traffic was clear to the left. The commander of the 747 then remarked, 'OK, what do you want now?' The controller replied by clearing the 747 to turn on to 330 and climb to 3,500ft. Under approach control, the 747 was radar vectored for a second ILS approach. It landed without further incident at 07.28hrs. It was after their return to Manchester that the One Eleven crew heard that the 747 had come too close for comfort during take-off.

Subsequent examinations of surface movement radar information at Manchester showed that the B747 was 1.5nm from the runway threshold as the A320 vacated the runway. As the One-Eleven began its take-off roll, the 747 was 0.5nm from the runway threshold. The 747 crossed the threshold as the One-Eleven passed link 'H', a separation distance of 850m. As the 747 crossed the touchdown zone, G-AYOP was halfway between link 'G' and link 'F', giving an approximate separation distance of 860m. As the 747 passed abeam link 'G', the One-Eleven was abeam link 'KC', an approximate separation distance of 480m. Then as the B747 passed abeam 'KC', the One-Eleven was approaching link 'B', an approximate separation distance of 700m. Finally, the 747 turned away to the right as it passed abeam link 'B'.

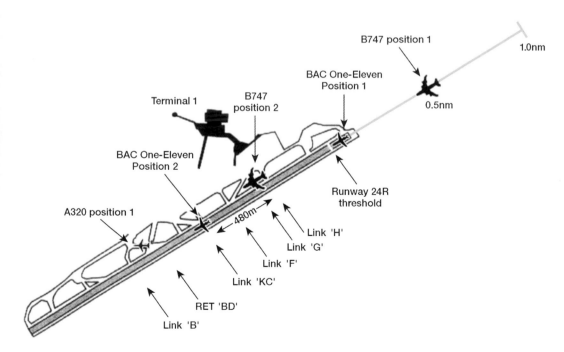

The Manual of Air Traffic Services states that 'unless specific procedures have been approved by the Civil Aviation Authority, a landing aircraft shall not be permitted to cross the beginning of the runway on its final approach until a preceding aircraft departing from the same runway is airborne'. Manchester had been given approval, subject to certain conditions, to use a 'Land After Departure Procedure'. The 'Land After Departure Procedure' for big jet aircraft in fine daylight weather states that:

'When the runway is temporarily occupied by departing traffic, landing clearance may be issued to an arriving aircraft provided that, at the time the landing aircraft crosses the landing threshold of the runway in use, the following minimum separation distances will exist:
'Either the departing aircraft will be airborne and at least 2,000 metres from the landing threshold, or the departing aircraft will not be airborne but will be at least 2,400 metres from the landing threshold.
'*The Air Controller is responsible for assessing that the specified separation minima will be achieved. The final responsibility to accept the landing remains with the pilot.*'

In an ideal world, strict criteria would be laid down and both controllers and pilots would know exactly what was and was not safe operating practice. But aviation it is not a nice, clear-cut business. Aircraft not only come into airports in many shapes and sizes with widely differing performance capabilities but also arrive in peaks and troughs rather than in a nice steady progression throughout the day. In a commercial world where passengers resent hanging around, there is strong pressure to push as many aircraft through an airport and quickly as possible. Consequently, Manchester International's 'Best Practice' to be applied to the 'Land After Departure Procedure' largely left final decisions to the judgement and experience of skilled professionals.

'When a "Land after Departure" clearance is being considered, a decision must be made that the required separation will be achieved. If it will not be achieved, then missed approach instructions must be issued and consideration given to stopping the departing aircraft. The point at which this decision should be finalised is around 2nm from touchdown, after taking into account all the relevant factors. A missed approach from this range, with an aircraft just commencing its departure roll, is much easier to resolve at this stage than later on short finals.'

Immediately after the incident, the controller concerned was withdrawn from duty as a precaution in case the incident affected the rest of his performance that day. No one was hurt in the incident, and the aim of the subsequent investigation was not to hang anyone out to dry but rather to see why the Best Practice guidelines had been insufficient in themselves.

Left:
Schematic of aircraft positions relative to Runway 24R at Manchester Airport on 16 September 2000.

The controller's apparent plan of action was to utilise the 'land after' procedure and he persevered with it despite a number of cues which should have alerted him to the fact that the plan was unworkable. The A320 was slow to vacate the runway, the One-Eleven was rolling slowly, and the incoming 747 was only 1.5 miles out when he cleared the One-Eleven for take-off. All these factors eroded the time available and militated against the plan working.

Looking back, even though he realised that the situation was 'tight', the controller reported experiencing some distortion in his perception of time. He reported thinking that the One-Eleven was 'taking an age' to roll down the runway, while the oncoming B747 was 'surprisingly fast'. Such distortion is not uncommon in time-dependent situations, and it often occurs when insufficient time has been allocated to carrying out a predetermined plan. Even after realising that the situation was becoming more critical, the controller stuck with his plan of action until control of the situation was effectively taken out of his hands by the 747 pilot's decision to trust in his four Pratt & Whitney PW4056 turbofans to get him out of trouble.

Past incidents have shown that, in such situations, individuals find it very difficult mentally to 'stand back', reassess their plans and make any necessary adjustments. There are various reasons for this. Sometimes people find it difficult to give up the comfort and reassurance of a plan, once formed. Sometimes they are already working to their full capacity and have few resources left to re-evaluate and amend existing plans. At other times they may be concerned about a perceived loss of face associated with having to abandon their chosen course of action.

This incident had all the hallmarks of a controller operating under stress. While being long on the human dimensions of the incident, the subsequent investigation failed to address the organisational and commercial pressures that can drive aviation professionals to make errors in the first place.

*   *   *   *   *

Back in 1944, Winston Churchill's Government approved the expansion of a military strip around the Middlesex hamlet of Heath Row to serve as a great dispatch point for the thousands of military personnel it was thought would be necessary to complete the war against the Japanese in the Far East. Wartime necessity then gave way to civilian convenience as the Heathrow strip was decked out with tents, walkways, toilets and a W. H. Smith newsagent's for its first fare-paying passengers in 1946.

Heathrow was originally set to have three sets of parallel runways laid out in a Star of David design with arrival and departure terminals in the middle. But as the airport became more and more popular, terminals and their associated infrastructure mushroomed out over former runways, leaving Heathrow flight operations centred on two main runways best suited to the prevailing winds — 27L/09R and 27R/09L, separated by approximately 1,340m — and one subsidiary, single-direction Runway 23.

Left:
BAC One-Eleven
G–AYOP at
Stansted in 1999.
*The Aviation
Picture Library*

Heathrow is rightly proud of being Europe's premier hub, but the downside of handling 65 million passengers a year is that its east-west runways are constantly used to maximum capacity. In an ideal world, one runway would be used for take-offs and its twin for landings to spread noise and environmental disturbance. However, up to six landing aircraft per hour can be accepted on a departure runway when the inbound holding delay is anticipated to be 20 minutes or greater. It is also best to land 'heavy' aircraft on the departure runway to reduce the impact of wake vortices on smaller aircraft close behind.

On 28 April 2000, Heathrow was using Runway 09 Right (09R) for take-off and Runway 09 Left (09L) for landing. The Heathrow Visual Control Room is manned by a supervisor, five air traffic controllers and various supporting staff. Three of the controllers deal with ground movements and two, 'Air Arrivals' and 'Air Departures', are responsible for all movements on the respective arrival and departure runways. The controllers' positions are on a raised dais facing east or west depending on the runway direction in use. This allows a good view of the runways and the approach and departure tracks.

In 'Air Departures', a 28-year-old ATC trainee was controlling take-offs from Runway 09R under the supervision of a 35-year-old mentor. At 13.55hrs, the Heathrow Intermediate Director (North) at the London Area Terminal Control Centre (LATCC) was given permission for 'Speedbird Six', a British Airways B747-436 (G-BNLY) out of Tokyo Narita, to land on 09R. At Heathrow, visual controllers are provided with individual Aerodrome Traffic Monitors, which display radar data to help them achieve maximum runway utilisation and aerodrome capacity. When Speedbird Six showed on the monitor screen, the mentor asked the trainee to assess the distance to touchdown. After some discussion, the two controllers agreed that the 747 was approximately 15nm out, which the trainee estimated equated to around five minutes' flying time.

As befits Europe's busiest airport, Heathrow always has a line of aircraft edging forward at the take-off holding point. Just after lunch on 28 April, there were a number of aircraft keen to get airborne because of earlier fog. The minimum Heathrow departure interval is one minute, but this interval can be varied depending on aircraft type and routeing. On this occasion, the trainee was allowing a two-minute departure interval between two of the aircraft in the departure sequence. The trainee estimated that it would take about six minutes for the aircraft cleared for conditional line-ups to get airborne, and she 'cocked' the last aircraft's Flight Progress Strip — a British Midland Airways Airbus A321, callsign 'Midland One November Zulu' — on the display as a reminder that that clearance might have to be cancelled.

The Heathrow Final Director is responsible for integrating aircraft inbound from the north and south and for vectoring and sequencing them for final approach. The Final Director normally retains control of inbound aircraft until the pilot reports that he is established on the ILS, when the flight is transferred to 'Air Arrivals' for landing on the arrivals runway, or to 'Air Departures' if the landing is to be made on the departure runway.

Heathrow Director informed Speedbird Six at 14.02hrs that G-BNLY was 'number one for nine right'. A minute later, Speedbird Six checked in with 'on frequency, range six miles, nine right'. The trainee responded, 'Speedbird Six,

continue approach, wind zero seven zero, eight knots.' With Speedbird Six less than six miles out and closing at around three miles a minute, three outbound aircraft still had provisional line-up clearances from the full-length threshold.

The trainee cleared the first, 'Shamrock Seven One Five', for take-off at 14.03:40. Around this time, the mentor advised the trainee to ensure that the next two aircraft for departure would be ready for immediate take-offs. At 14.04:20, with 'Lufthansa Four Five Seven Seven' lining up on the runway, the trainee transmitted to the third aircraft, 'Midland One November Zulu, when so cleared can you take an immediate departure, there's landing traffic four miles.' When Midland One November Zulu confirmed that he could, the trainee cleared the Lufthansa 737 for an immediate take-off at 14.04:40. The Lufthansa jet replied, 'Rolling.'

By now, British Midland Airways Airbus G-MIDF was taxying towards the runway. The crew could see a return on their Traffic Alert & Collision Avoidance System (TCAS) display indicating an aircraft at approximately two miles on final approach to Runway 09R. The crew asked, 'Midland One November Zulu, just confirm we are cleared to line up?' The trainee immediately replied in the affirmative and Midland One November Zulu acknowledged this message at 14.04:50. The trainee then gave a conditional line-up clearance 'after landing Seven Four Seven at two miles' to the next aircraft in the departure sequence. This was acknowledged at 14.05:10.

At this point, the mentor transmitted to the aircraft on approach, 'Speedbird Six, keep it coming, there's

Above:
British Airways
747-436 G-BNLY.
*The Aviation
Picture Library*

131

one to roll, the wind zero seven zero, eight knots.' After Speedbird Six acknowledged this message, the mentor transmitted: 'Midland One November Zulu, start powering up on the brakes and you're clear immediate take-off zero nine right, the wind zero six zero, eight knots.' The crew acknowledged, but the mentor transmitted within 10 seconds, 'Midland One November Zulu, cancel take-off, I say cancel again take-off, hold position.' This instruction was acknowledged with, 'Holding position, One November Zulu.' The mentor then transmitted, 'Speedbird Six go around, say again, go around, acknowledge.' Speedbird Six immediately acknowledged this instruction, though that must have been the last thing the 381 passengers and crew on board the 747 would have wanted after their long haul from Japan.

The mentor was aware that his priorities now were to deconflict Speedbird Six on its go-around from the Lufthansa 737 that had just got airborne. He transferred Shamrock Seven One Five to his departure frequency and then checked the altitude of the Lufthansa jet before restricting Speedbird Six initially to an altitude of 2,000ft. The required separation was maintained between the two aircraft and they were subsequently transferred to other control frequencies. Speedbird Six was then vectored back for an uneventful approach. G-BNLY landed on Runway 09R at 14.18hrs.

From their position on the runway, the crew of the British Midland Airbus were 'startled to see an aircraft flying directly above them', which was probably an understatement. Midland One November Zulu was cleared for take-off at 14.08hrs and transferred to a departure frequency just over a minute later. Both mentor and trainee were relieved from duty approximately five minutes after the incident.

Both controllers submitted a report immediately after the incident and were interviewed some 10 days later. They both considered that they were adequately rested and that all relevant ATC equipment was serviceable. The mentor had come on duty slightly early and had not realised that he had been allocated a training role during his shift. He assumed his duties as the Air Departures controller at 13.20hrs and had been operating for some 10 minutes when the trainee arrived and told him that he was programmed to train her during the shift. Since he had trained her some five to six weeks previously, he was content to take the duty and asked her to sit and monitor while he updated her on the situation. He was aware that she had completed approximately one third of her training and that she was progressing well. Once she was familiar with the situation, he moved out of his seat and allowed her to take control under his supervision.

As he monitored her, he considered that she was controlling in a confident manner and was making good decisions. He heard the Intermediate Director (North) call about 'Speedbird Six', and he allowed her to make the decision to accept the 747 for a landing on Runway 09R without any input from him. The mentor commented that he would not personally have accepted it, but that he was perfectly happy that it could be safely integrated. His recollection was that there were still about four aircraft to get airborne.

At about this time, the mentor became aware that Shamrock Seven One Five had still not started its take-off roll, and he realised that the Irish airliner required a two-minute separation to comply with spacing regulations. He still

considered that Speedbird Six could be safely integrated but they had to act without undue delay. He instructed the trainee to confirm with the next two departing aircraft that they were ready for immediate take-off. She confirmed this with the Midland crew and then cleared Lufthansa Four Five Seven Seven for an immediate take-off. Shortly afterwards, the mentor heard the British Midland crew question their line-up clearance, but he considered that the Airbus was already past the 'Cat 1 Hold' and so did not interject when the trainee confirmed the line-up clearance. The mentor was now aware of the rapidly developing situation. He saw Speedbird Six on finals and took control of the radio. He instructed Speedbird Six to continue and then cleared Midland One November Zulu for an immediate take-off. However, he realised almost immediately that this plan was flawed. He cancelled the Airbus's take-off clearance and ordered Speedbird Six to go around. His next priority was to ensure that Speedbird Six was deconflicted with Lufthansa Four Five Seven Seven, and this he did primarily by vertical separation.

The trainee believed that, at the time of the discussion about how far Speedbird Six was out over Berkshire, she still had at least five aircraft conditionally cleared to line up. She mentally calculated that Speedbird Six would land in five minutes and that it would take six minutes for the departing aircraft to get airborne. Her recollection was that she asked the mentor

Above:
British Midland
A-321 G-MIDF
at Heathrow in
February 2000.
*The Aviation
Picture Library*

if she should cancel the line-up clearance for the British Midland A321 as she 'cocked' the relevant Flight Progress Strip (FPS). He replied that she should continue and see how things progressed. Then, when Speedbird Six checked in on her frequency, she did not inform the crew of the number of aircraft still to get airborne because she expected the number to change. Shortly afterwards, she cleared Shamrock Seven One Five for take-off and again asked the mentor if she should cancel the line-up clearance for Midland One November Zulu. He told her to check if the two aircraft were ready for immediate take-off. She recalled that, at the time when Speedbird Six was at two miles, she had noted this range from the Air Traffic Monitor but also she saw the lights of the aircraft.

Midland One November Zulu had moved to the right to allow Lufthansa Four Five Seven Seven to overtake and line up on the full length of Runway 09R. In response to the controller, the crew confirmed that they were ready for an immediate take-off. Then, after the Lufthansa 737 had acknowledged its take-off clearance, the crew of Midland One November Zulu saw an aircraft indicating on TCAS at about two miles range from their position, though they did not have visual contact with the 747. The Midland commander was sure that his A321 was still short of the 'Cat 1 Hold' when he queried his line-up clearance. He reacted immediately when this clearance was confirmed and was fully prepared for take-off. On the runway, he was somewhat surprised by a non-standard call instructing him to 'start powering up on the brakes' but the first officer, as handling pilot, increased power to the normal stable position in preparation for take-off. Once he had received the take-off clearance, the first officer released the brakes but almost immediately re-applied them as clearance was cancelled. Both pilots estimated that their aircraft moved only a few metres before stopping. Thereafter, both pilots were startled to see an aircraft fly directly over them, and the commander transmitted to the controller that he would be filing a report.

Speedbird Six had flown from Narita Airport in Japan and made initial contact with the Heathrow Intermediate Director (North) at 13.49hrs. During the initial approach, the commander operated as handling pilot and, in accordance with normal company procedures, was prepared for the first officer to assume these duties when he (the first officer) became visual with the runway. Both pilots recalled that they were transferred to 'Heathrow Tower' at approximately 8km range from touchdown. The first officer became visual with the runway at about 900ft above ground level, and took the handling duties, with the aircraft fully configured for landing and fully established on the ILS. As he did so, he could see the runway but could not make out any aircraft on it. The commander looked up and also saw the runway. Neither recalled hearing any ATC information about the number of aircraft still to get airborne but, as they continued their approach, they both had a mental picture of one aircraft still to take off. The commander then saw an aircraft, which seemed to be lined up at an intersection on Runway 09R (close to the displaced runway threshold), and understood that to be the relevant aircraft. He expected landing clearance once that aircraft lifted off, and his impression was reinforced when the controller called, 'Keep it coming, there's one to roll.' However, as Speedbird Six approached about 200ft above ground, the commander saw another aircraft lined up on the runway before the displaced threshold. The commander later commented that this aircraft was difficult to see against the runway surface. As the crew initiated a go-around, the controller also called for them to overshoot. The commander considered that the first officer carried out an immediate and positive go-around with minimal height loss. Thereafter, they were vectored back for an uneventful landing on Runway 09R. There were neither injuries to anyone on board nor damage to the aircraft.

*   *   *   *   *

At the time of the incident, visibility at Heathrow was 6km, cloud was overcast at 900ft and the surface wind was 030°/06kt. The flight crew on both aircraft were qualified and current to operate the flights.

Subsequent analysis of recorded data show that G-BNLY maintained an accurate ILS approach to Runway 09R. When the crew were advised by ATC to 'keep it coming', the aircraft was at 450ft above ground. Twenty seconds later, when the crew were instructed to 'go around', the 747 was passing 175ft. Subsequent height loss after the initiation of the go-around was about 50ft, and it was calculated that the minimum height reached by G-BNLY was 118ft above ground.

Underneath, British Midland Airbus G-MIDF sported the standard company colour scheme of dark blue and grey on the upper surfaces. The normal navigation and beacon lights were serviceable and on. The beacon lights were red and flashing, and one was located on top of the fuselage. The strobe lights were selected to automatic mode, which meant that they illuminated in concert with white flashing wingtip and tail cone lights only when the aircraft was airborne. As the tail fin of G-MIDF extended 38ft 7in above the ground, the 747 with 381 people on board could have missed the Airbus by as little as 80ft, which was too close for comfort in anybody's book.

Radio contact was established between 'Air Departures' and Midland One November Zulu at 13.55hrs when the crew were instructed to hold on the right-hand side of the holding area at the threshold of Runway 09R. At 13.57hrs the crew were instructed to line up after Lufthansa Four Five Seven Seven which would pass them on their left side. At this time, there were five aircraft to depart ahead of Midland One November Zulu; additionally, during these departures, two further aircraft were co-ordinated and cleared to taxi across Runway 09R.

The full-length take-off distance for 09R is 3,658m but, due to a displaced threshold 305m from the beginning of the runway, the landing distance is 3,353m. The rules for take-off on such a runway are clearly laid down in the Manual of Air Traffic Services:

> 'An aircraft shall not be permitted to begin take-off until the preceding departing aircraft is seen to be airborne or has reported "airborne" by radio and all preceding landing aircraft have vacated the runway in use. 'A landing aircraft will not be permitted to cross the beginning of the runway on its final approach until a preceding aircraft is airborne.'

However, Heathrow ATC has dispensation from the Civil Aviation Authority to use the following 'After the Departing Procedure':

'When the runway-in-use is temporarily occupied by other traffic, landing clearance may be issued to an arriving aircraft provided that the Air Controller is satisfied that at the time the aircraft crosses the threshold of the runway-in-use . . . the departing aircraft will be airborne and at least 2,000m from the threshold of the runway-in-use, or if not airborne, at least 2,500m from the threshold of the runway-in-use.'

For this procedure to be used — and it was well practised at LHR — the reported meteorological conditions have to be equal to or better than 6km visibility and a cloud ceiling of 1,000ft. At the time of the incident, the reported cloud ceiling was 900ft. Therefore, 'Air Departures' had to ensure that Midland One November Zulu was airborne before clearing Speedbird Six to land.

At the time of the incident, the workload on 'Air Departures' was assessed by the controllers as medium and the acceptance, by the trainee, of Speedbird Six for a landing on Runway 09R was reasonable. The fact that the mentor subsequently stated that he would not have accepted the aircraft was not related to workload or safety, and indeed he felt no need to countermand the trainee's decision. Therefore, the acceptance of the aircraft was a reasonable decision and should not have resulted in the subsequent incident.

With some five minutes to go before Speedbird Six crossed the runway threshold, it was reasonable for the controllers to see how the situation unfolded. The first point at which warning bells should have started to ring was when Speedbird Six checked in with 'Air Departures' at six miles (about two minutes) to go before touchdown. With three aircraft still to get airborne, and a one-minute interval between each of them, the existing plan was no longer viable. Thereafter, more warning cues should have flashed up, including a prompt by Midland One November Zulu. Finally, when action was taken, the wrong option was initially adopted.

Who could or should have prevented the near-disaster? The Speedbird Six crew were vectored on to the ILS at 180kt and told that they were 'Number one for nine right'. Although they knew that there was no landing traffic to impede their progress, experience would have taught them that there would be departing aircraft. Once the crew checked in with 'Air Departures', they were not given any information on the number of aircraft still to take off. If this message had been passed, it might have helped the crew to formulate an accurate mental picture, especially as this was the first time that they were on a common frequency with the departing aircraft. These departing aircraft had already been given line-up clearance, so the Speedbird crew would only have been able to assess the full situation by subsequent radio transmissions during the 150 seconds between check-in and commencing their go-around. Radio transcripts confirmed that three different aircraft were given clearance to take off after Speedbird Six checked in. Additionally, the call of 'Speedbird Six, keep it coming, there's one to roll' was a big hint.

The 747 crew believed that this preceding aircraft was the Lufthansa 737 near the intersection, so it was understandable that they focused on the progress of this aircraft while awaiting their landing clearance. As British Midland company policy laid down that strobe lights should be on automatic, the fact that the piercing strobe lights were not yet active denied Speedbird Six an additional alert

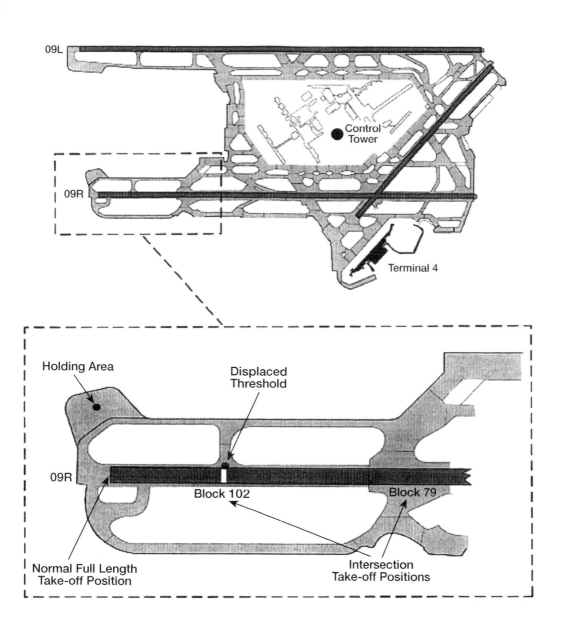

09L

Control
Tower

Terminal 4

Holding Area

Displaced
Threshold

09R

Block 102

Block 79

Normal Full Length
Take-off Position

Intersection
Take-off Positions

Above:
London Heathrow
airport chart,
showing the
displaced 09R
threshold.

feature. The 747 crew became aware that there was another aircraft on the runway only about the time the controller instructed them to 'go around'.

In sum, the crew of Speedbird Six, tired after a long haul from the Far East and arriving at Europe's busiest airport in the midst of an almost never-ending rush hour, were not given the information they needed for a full mental picture of what was happening below them. Turning to the Midland One November Zulu crew, they were aware of their position in the line-up queue and that an aircraft was approaching to land on Runway 09R. The commander of the Midland A321 had been asked if he was ready for an immediate take-off and had replied in the affirmative. But although the Midland crew were fully prepared for departure, just after Lufthansa Four Five Seven Seven started its take-off roll, the Airbus commander noted from his TCAS display that an aircraft was on final approach at about two miles range. Although he did not acquire Speedbird Six visually, he questioned the line-up clearance only for the controller to reply immediately, 'Affirm'. The Midland crew's subsequent reaction to this message was prompt. Although the crew could have refused the line-up clearance, they would have done so only if they had considered that acceptance of the clearance was unsafe.

At Heathrow, crews are operating in a busy environment, with well-respected controllers and against a backdrop of intensive radio traffic. At the time, Speedbird Six had not been cleared to land, the crew of Midland One November Zulu had no visual contact with the approaching aircraft and the commander had received an immediate reply to his query. Therefore, the action of the Midland crew in entering the runway was reasonable. The commander was then surprised to hear the instruction to 'Start powering up on the brakes,' but the message continued with a clearance to take off, so the crew continued with their normal procedure. They also reacted promptly to the call to 'Hold position'.

Thereafter, the Midland commander tried twice to pass a message but the controller was concentrating on deconflicting Speedbird Six and Lufthansa Four Five Seven Seven. The commander of Midland One November Zulu then advised the controller that he would be filing a Mandatory Occurrence Report, as he considered that the incident had been 'very dangerous indeed'. The A321 commander subsequently acknowledged that the latter part of his message was inappropriate, but the crew had just been startled by the close overflight of Speedbird Six and the message content was an understandable reaction to the event. Nevertheless, it could have adversely affected the controller and diverted him away from his primary task of deconflicting two aircraft.

The use of strobe lights would have made the Airbus more conspicuous on the runway, and some airlines have operating procedures requiring their crews to activate strobe lights whenever their aircraft are on an active runway. However, there is no national regulation requiring this. The use of dazzling and piercing strobe lights on the ground could be disturbing to other crews awaiting take-off, particularly at night. Nevertheless, the use of strobe lights would increase the conspicuity of aircraft on an active runway.

What of the ATC team of mentor and trainee? Not all controllers enjoy acting as an On the Job Training Instructor, and the mentor in this incident was one such individual. He confirmed that he had not sought the responsibility but

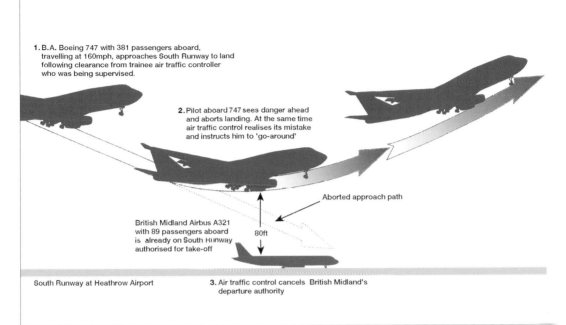

1. B.A. Boeing 747 with 381 passengers aboard, travelling at 160mph, approaches South Runway to land following clearance from trainee air traffic controller who was being supervised.

2. Pilot aboard 747 sees danger ahead and aborts landing. At the same time air traffic control realises its mistake and instructs him to 'go-around'

Aborted approach path

British Midland Airbus A321 with 89 passengers aboard is already on South Runway authorised for take-off

80ft

South Runway at Heathrow Airport

3. Air traffic control cancels British Midland's departure authority

accepted it as part of his function at the unit. This, coupled with the fact that he had not expected to be acting as mentor on the day in question, was unlikely, at least initially, to have placed him in the best frame of mind to fulfil the training function, although the trainee did not detect anything unusual in the mentor's attitude toward her or the training on the day in question.

The mentor was aware that his trainee was relatively inexperienced, especially on the Air Departures position. However, reports indicate that this particular trainee was progressing well but, since she had completed only about one third of her training at the time of the incident, it would be reasonable to expect that she would be receiving not only the mentor's close attention but also some advice with regard to making decisions and formulating plans. When the request was received for landing on Runway 09R, the mentor reported that the trainee had looked questioningly at him, presumably for a decision. He states that, in response, he 'blanked her out', leaving the trainee to make the decision for herself. He then went on to discuss the fact that he would not have accepted the landing aircraft, though there was no good safety reason for not doing so. This was the first occasion during training that the trainee had been required to make such a decision herself. On previous similar occasions, including sessions with the same mentor, she had been acting on decisions made for her. The trainee was placed in a somewhat ambiguous situation. She had sought advice, been refused it and

Above:
Diagram showing one of the most dangerous near-collisions in British aviation history

then, having had to take the decision, had been in the position of having that decision criticised during the operational session. Nevertheless, once Speedbird Six was accepted for landing on 09R, it was the mentor's responsibility to assist the trainee in ensuring safe integration of arriving with departing traffic.

When Speedbird Six was accepted, five aircraft had been given conditional line-up clearances on Runway 09R. Subsequently, when the 747 was about 15nm from touchdown and a discussion about aircraft spacing took place between the mentor and his trainee, two of these aircraft had departed and a further two (Lufthansa Four Five Seven Seven and Midland One November Zulu) had received conditional line-up clearances. The mentor believed that the departure order was such that only one minute was required between departures and in his opinion, although the situation was tight, the plan would work. However, he was unaware that the order implemented by his trainee required a two-minute spacing between two of the departing flights. It is assessed that this oversight occurred because he was not monitoring his trainee's actions closely enough.

What of the Heathrow organisation itself? Co-ordination for the use of the departure runway by Speedbird Six was in accordance with standard operating procedures. There were no safety reasons to prevent the use of the departure runway for arriving aircraft, but, to comply with the clearance to land procedures with a 900ft cloud base, Midland One November Zulu had to be airborne before Speedbird Six could be cleared to land. There was insufficient spacing between these flights to meet this requirement.

To ensure maximum runway utilisation in accordance with ATC regulations, or to cram as many airliners in as possible to maximise revenue, depending on your point of view, Heathrow and many other airports make regular use of conditional line-up clearances for a number of aircraft at a time. However, this incident illustrates the risks inherent in adopting this technique for a large number of aircraft when the runway is being used for landing and take-off. Heathrow ATC Operations staff are currently considering the introduction of a limitation on the number of aircraft which can hold such conditional clearances at any one time.

The correct procedure, whereby the arriving aircraft was given a go-around and the departing flight's take-off clearance was cancelled, was carried out, albeit at a very late stage and without the use of standard phraseology. However, although incorrect phraseology was used, it was not considered that this had any effect on the incident, as the pilot knew what was expected of him and reacted accordingly.

The official investigation into the close call of 28 April concluded that the ATC mentor allowed the situation to develop to the point where Speedbird Six could not be safely integrated with the departure of Midland One November Zulu. He was transferred to a 'less busy' airport. No blame was attached to the trainee or anyone else.

It is hard to argue with this assessment in narrow legal terms, but it is the soft option to let one man carry the can while making a few vacuous recommendations about the use of strobe lights, the need for formal briefings of student controllers and the importance of regularising the selection and monitoring of On the Job Air Traffic Control Training Instructors. Heathrow controllers enjoy a well-deserved reputation for very high standards of controlling, but the best humans can cope with only so much pressure, and if air traffic controllers are pushed to the limits for too long, too often, they tend to focus on just one aspect, and full situational awareness goes out of the window.

In no particular order, others shared degrees of responsibility with the unfortunate controlling mentor. The 747 commander had 23,000 flying hours of which 3,800 were on type, which meant that he must have flown into crowded Heathrow countless times. He and his first officer ought to have picked out the British Midland Airbus on the threshold ahead before approaching their relatively low decision height of 200ft, or what is the aircrew eyesight test for?

Then there are the planners. Heathrow has four terminals, and the public inquiry into Terminal 4 back in 1979 set a limit of 260,000 flights a year. In 1992 the British Airports Authority stated that Heathrow was 'full' with 410,000 annual take-offs and landings. The Authority sought approval for a fifth terminal at Heathrow on the grounds that movements would rise to 453,000 by 2016. No new runway or terminal has been built at Heathrow in the meantime, but total movements in 2000 stood at 457,000.

During April 2000 there were 19,268 arrivals and 19,275 departures (including helicopters) at Heathrow. Runway 09L had 8,693 landings and no departures, while 09R had 492 arrivals in and among 9,413 departures. Including the incident on 28 April, there were 31 go-rounds that month. The Civil Aviation Authority now admits that there is one emergency each day at Heathrow, and a 747 pilot has gone on record as saying that the Heathrow approach is now so busy that 'there is no margin for error and the hairs stand up on the back of my neck'. Safety is everyone's business, and until London gets another airport or Heathrow gets a third runway, the powers that be should stop trying to stuff more and more airliners into Heathrow.

11

Unhappy Landings

R AF Northolt is in Middlesex, to the west of London. Northolt made its name as a key fighter base in the defence of London during the Battle of Britain, and after the war it became London's main airport until the new site at Heathrow, positioned about 5nm to the south, was completed. Thereafter, Northolt continued in business both as an RAF transport base and as a place where VIP traffic could arrive in London discreetly. Princess Diana's body was flown back from Paris to Northolt in 1997, and the airfield still houses the aircraft of No 32 (Royal) Squadron, formerly known as the Queen's Flight.

Northolt's prefabricated terminal and operations buildings were hurriedly erected after the war, and many of them stood the test of time until the construction of the new operations building in the mid-1990s. A ramp was built on the south side, and the main 07/25 runway was resurfaced. Runway 25, with an available landing distance of 5,525ft/1,684m, was the in-use runway on 13 August 1996 when a Spanish-registered Gates Learjet operated by MAC Aviation Sociedad Anonima came in for its final landing.

The designated flight crew took off from their home base at Zaragoza, northeast Spain, at approximately 04.20hrs on 13 August. Once airborne, the crew discovered a fault in their aircraft's directional gyro (DG) system, and as they knew that they would have to fly a Precision Approach Radar (PAR) into RAF Northolt — there is no ILS installed at Northolt — the crew returned to Zaragoza to change aircraft. The flight finally departed Zaragoza at 05.25hrs in Learjet 25B EC-CKR and flew to Mallorca in the Balearic Islands.

At Palma de Mallorca the crew refuelled the aircraft and, after their one female passenger had boarded, they departed at approximately 06.45hrs on an IFR flight plan for RAF Northolt. The flight to the UK, callsign 'Mike Alfa Quebec One Two Three', which was flown at FL390, passed uneventfully.

On approaching the UK, the 23-year-old Learjet was descended under radar control to transit to RAF Northolt via the Biggin Hill VOR beacon. London Heathrow has a large, almost rectangular control zone around it, and Northolt lies firmly within that zone.

Left:
Learjet 25 EC-CKR sitting all forlorn on the A40 carriageway after running off the end of the RAF Northolt runway.
The Aviation Picture Library

Given the need to co-ordinate all Northolt traffic with that of its much bigger and busier neighbour, Heathrow controlled much of what happened in the Northolt pattern, down to deciding which Northolt runway was to be in use.

Approaching the UK Flight Information Region (FIR) boundary, the crew contacted the London Area Terminal Control Centre (LATCC) at 08.25hrs and were given routine clearances to position the aircraft for its approach to Northolt. MAQ123's descent continued normally and control was passed to the Biggin sector of LATCC and then via Heathrow Director South and Heathrow Director North to the RAF Northolt Director.

Handover from the Heathrow Director was effected with EC-CKR descending from FL70 to 4,000ft on a heading of 360°. Northolt Director then cleared MAQ123 to continue its descent to 3,000ft and passed the Northolt weather of 340°/9kt surface wind, 12km visibility, nil weather, cloud one octa at 1,500ft, five octas at 2,200ft, temperature +16°C and QFE 1010mb. As there is high ground to the east of Northolt airfield, approaches to Runway 25 use a 3.5° glide path. So the next message instructed the Learjet pilot to turn downwind on a heading of 070°, and he was advised that the decision altitude was 330ft, the field elevation was 124ft and the radar talkdown procedure was to be based on a 3.5° glide path to Runway 25.

As priority traffic was scheduled to depart Northolt around that time, the Learjet was extended downwind to a distance of 10nm before being turned onto a heading of 160°M. This was followed shortly afterwards by an instruction to turn onto 250°M and descend to 1,800ft to intercept the final approach. At just over nine miles from the airfield, MAQ123 was identified by the PAR controller and given a further turn 10° starboard. At 08.55hrs and at a distance from the runway of 4½ miles, the Learjet pilot was instructed to begin the final descent. The first officer subsequently confirmed that the pilots had visual contact with the runway from this time onwards. However, the commander decided that he would fly the approach solely on instruments until the decision height in order to obtain maximum training value.

Initially, the aircraft was slightly high in the descent but this was corrected and at 3.5nm it was on the glide slope. At this point the pilot was asked to confirm that his landing gear was down and locked, as is normal procedure at Northolt. This request was repeated three times using the following words, 'Three and a half miles, check gear, acknowledge', followed by 'One Two Three, confirm gear down' and then 'One Two Three, confirm undercarriage is down'. MAQ123 then replied, 'Affirmative sir, gear is down and locked, One Two Three.' During this exchange the Learjet was seen to go above the glide path. At 2.5nm, landing clearance was confirmed and the aircraft was advised of the surface wind. The next call from the controller was that MAQ123 was 'Above glide path, one and a half miles, turn right five degrees heading two six five, slightly left of centreline, above glide path, correcting nicely . . . Tail wind of four knots, slightly left of centreline, slightly above glide path correcting nicely, one mile.' At decision altitude, which was half a mile from the runway, the aircraft was still above the glide path though it was seen to be correcting towards it.

On arrival, at 08.57hrs, the Learjet was observed to be higher than normal at the runway threshold and to land beyond the normal touchdown point. The length of runway remaining at the actual point of touchdown was estimated at 3,125ft (952m). Towards the end of the landing roll EC-CKR veered to the right and then swerved to the left before overrunning the end of the runway. The Learjet continued in a southwesterly direction towards the airfield boundary. After bursting through a high chain-link boundary fence, the business jet crossed a shallow kerb on to the A40 trunk road, which runs alongside the southern perimeter of the airfield.

As the aircraft motored across the three-lane eastbound carriageway of one of the busiest routes into London, it collided with a Ford Transit van travelling at 60-70mph in the middle of the three lanes. The van had snuck under the leading edge of the right wing without striking the tip tank but then it impacted with the Learjet fuselage immediately in front of the wing leading edge; this caused the aircraft to yaw left and severed the aircraft nose. EC-CKR came to rest in the left-hand lane of the road, with the van embedded in the right side of its fuselage immediately forward of the right wing.

EC-CKR's final approach and landing were observed by the Local Controller and his assistant from the ATC tower. Both of them were concerned by the abnormally high speed on final approach and late touchdown, and the landing roll was followed closely by the assistant controller, using binoculars. The controller activated the crash alarm as the aircraft left the runway, and at the same time his assistant telephoned the civil emergency services. The rapid reaction of the Local Controller and his assistant ensured that the civil police and ambulance service arrived at the scene within five minutes of the accident occurring.

The prompt reaction of the airfield fire service averted the possibility of a post-impact fire, which was fortunate because both the first officer and the van driver were trapped in their seats. The impact of the van was severe, but it was mainly forward of the rearmost passenger seats and aft of the flightdeck and none of the seats in that area was occupied. Although the van driver was not seriously injured, it was some time before he could be freed from the wreckage, due to extensive deformation of the front of the van, which was partly underneath the fuselage. There were no skid marks from the van, but a following vehicle had left skid marks which showed that it had swerved into the unoccupied right-hand lane to avoid impact.

As luck would have it, the accident was observed by the crew of an Air Ambulance helicopter that was holding 2nm south of Northolt awaiting clearance for an en-route crossing of the airfield. After receiving clearance to attend the accident, the helicopter crew landed some 50m from the Learjet. The helicopter co-pilot, together with a doctor and a paramedic who were also on board, attended the accident victims. RAF Northolt firefighters were rapidly on the scene and they applied foam to the wreckage that was by now leaking aviation fuel. Before reaching the road the aircraft had demolished three steel-reinforced concrete posts in the boundary fence and two low-level lamp standards. The power supply to the lamp standards remained live until the local authority disconnected it. The aircraft batteries were disconnected by the emergency services.

There was no fire and all the accident victims were taken by ambulance to nearby hospitals. The two flight crew and their passenger received minor injuries, but the first officer's concussion and bruising dictated a stay in hospital for two days. The van driver was treated for cuts and shock and was later discharged from hospital. The aircraft and the Transit van were destroyed.

*　　*　　*　　*　　*

Because the accident involved a civilian aircraft landing at a military airfield, the Ministers for Defence and Transport agreed that the investigation should be carried out by the AAIB. Under the provisions of the Convention on International Civil Aviation, a representative from the Spanish Directorate General of Civil Aviation participated in the investigation.

The aircraft wreckage was removed to the AAIB facility at Farnborough for detailed examination. Under existing Spanish airworthiness requirements, the Learjet, because of its weight category, was not required to be fitted with flight recorders but recordings of ATC exchanges and radar were available for analysis.

At the time of the accident, the Northolt weather comprised a surface wind of 360°/13kt with no gusts, 20km visibility, 2/8 cloud at 1,800ft, 6/8 at 2,500ft, temperature 17°C and QNH 1015mb. When the Learjet was less than two miles from the runway, the surface wind was 010°/15kt and shortly afterwards the pilots were told of a tail wind of 4kt. These were reasonable landing conditions.

Navigation during the letdown and approach for landing was conducted exclusively using the ground-based radar. Some 10 years previously the Ministry of Defence had considered the installation of an ILS at Northolt, but surrounding terrain prevented the installation of a standard 3° glide path and progress on the provision of ILS was halted.

Touchdown point for Runway 25, when using a PAR approach, was 700ft (213m) from the runway end, and therefore the landing distance available was 4,825ft (1,468m). The ground beyond the threshold of both ends of the runway was grassed and the overrun at the 07 end sloped gently, descending 3ft over a distance of 315ft (96m) to the boundary fence where the Learjet broke through onto the A40. The ideal international standard for civilian aerodromes is to position a strip 'before the threshold and beyond the end of the runway or stopway for a distance of at least 60m [197ft]' to act as an emergency overrun. No such strip was available at Northolt.

Learjet EC-CKR, with 4,596 hours on the clock, had been on the Spanish Civil Register from new. No relevant technical defects had arisen in the previous 30 days of operation, and there were no carried forward defects. On 13 August the aircraft was within its maximum certified landing limit, and it was correctly loaded within its centre of gravity limits. After the aircraft had been recovered to the AAIB facility at Farnborough, the pitot and static lines were leak checked and found satisfactory. The two ASIs were found to be within 2kt throughout the range 80-300kt.

Post-accident reconstruction indicated that the Learjet's final speed at touchdown was 158kt. Using data from Learjet's Airworthiness Department, AAIB investigators calculated that the normal speed for landing on Runway 25 at Northolt in the conditions pertaining on 13 August was 117kt. The following table shows the calculated distance in feet required to bring a Learjet weighing

146

9,951lb to a stop, assuming full application of brakes at different speeds with spoilers either deployed or stowed:

V full brake	158kt	148kt	127kt	117kt
Spoilers deployed	2,630ft	2,380ft	2,010ft	1,775ft
Spoilers stowed	5,630ft	4,630ft	3,380ft	2,750ft

As EC-CKR weighed 10,100lb, the actual landing distance required was 3,300ft. To stop in this distance, the Flight Manual specified that the following procedure must be applied:

(a) Approach through the 50ft point over the end of the runway at 1.3 times the stall speed with flaps and gear down.
(b) Approach using a glideslope of 2.5°.
(c) Spoilers to be extended immediately after touchdown.
(d) Wheel Brakes — Apply as soon as practical and continue braking action until the airplane stops.

The aircraft's touchdown point was calculated at between 2,708ft and 3,125ft (825m and 952m) from the A40 end of the runway, based on the eyewitness's evidence and allowing a margin for parallax error. It was clear from Learjet performance data that, provided the spoilers had been deployed, it should have been possible to bring the aircraft to a stop within the runway distance remaining even at the highest estimated touchdown speed of 158kt. However, without the use of spoilers, even at the lowest estimated touchdown speed of 148kt (158kt minus 10kt tolerance error), the landing roll required would be 4,630ft (1,411m) which was greater than the estimated distance remaining at the point of touchdown.

Examination of the normal and emergency brakes, and the spoiler and flap systems, showed them to be serviceable prior to the accident. When AAIB investigators arrived at the site just over an hour after the accident, the Learjet's gear was down and locked, though the right main leg had collapsed in the direction of retraction. Both flaps were in an almost fully extended position, touching the road surface, but the lift spoilers were found to be retracted and fully faired. Photographs of the aircraft taken within a few minutes of the accident showed both spoilers fully retracted. Examination of the patterns left on the wing by dried extinguisher foam showed that the spoilers were not extended when the aircraft was foamed.

The spoilers, like the flaps, were controlled from an electrical switch on the flightdeck. Mechanical locks in the spoiler actuators prevented the spoilers from extending unless hydraulic pressure was available. The spoilers could run back without hydraulic pressure, and the last part of the retraction movement caused the mechanical locks to re-engage as the spoilers faired with the wing, but this required additional forces to be applied to overcome the forces in the spoiler lock mechanisms.

Checks were conducted to determine the likelihood of the spoilers' retracting during or soon after the overrun. EC-CKR's spoiler system was rebuilt and refitted at Farnborough. The system was electrically energised, bled and function-tested satisfactorily. The spoilers were then extended fully using the hand pump, and the system shut down electrically and the pressure released at the pump. The spoilers remained fully extended 24 hours later. Several tests were then conducted to simulate the sequence of events during the overrun. In each test the system was electrically energised and the spoilers were hand pumped to the fully extended position. All these tests showed that the spoilers would retract only slowly and even then this was insufficient to allow them to re-enter the internal locks and fair fully with the wing profile.

The engineering investigation concluded that the serviceable spoilers were fully stowed when the aircraft came to rest. Investigators deemed it unlikely that this resulted from any specific action by the crew to stow the spoilers during the overrun. This assessment, plus the landing data, suggested that the spoilers were not selected during the landing.

Which leads to the human dimension of what happened on the flightdeck of EC-CKR. At the time of the accident both pilots were properly licensed, rested and in good health, and there was no evidence of any incapacitating illness or condition brought on by the use of either drugs or alcohol. The 39-year-old Learjet commander had learned to fly with the Spanish Air Force. The majority of his flying experience had been gained on the C-130 Hercules transport, of which 2,000 hours had been in command. On leaving the Air Force he joined MAC Aviation on 1 June 1990, and he converted onto the Learjet at Zaragoza. He had operated as a commander since joining MAC Aviation and had flown exclusively within Europe. He had flown 1,900 hours on type, and 39 in the previous 90 days. This was his first flight into RAF Northolt.

The first officer was 14 years older. He too had learned to fly in the Spanish Air Force, but his background had been on a variety of single-seat military fast jets. On leaving the military, he took up a post with the civil aviation authority, specialising in civil aircraft dispatch, where he remained for 15 years. He joined MAC Aviation in May 1989 and was converted onto the Learjet. He had operated exclusively as a first officer since joining and had declined an offer of a command on the grounds that he was not prepared to take on the extra responsibility.

Up to arrival over London, all recorded radio transmissions from MAQ123 were clear and concise. However, immediately after transfer of control to Northolt Talkdown, the non-handling pilot had difficulty in understanding some of the ATC phraseology. The initial transmission by the controller — 'Mike Alfa Quebec One Two Three, Northolt Talkdown, identified nine miles, read back QNH' — threw the first officer, who was operating the radios. Although there was nothing wrong with this transmission, it did not follow the standard civilian format in that it was not normal practice for a controller at this stage to ask a pilot to read back the QNH. This instruction, coupled with a somewhat staccato delivery, was not readily understood.

Notwithstanding his experience in military and civil aviation, the first officer showed during the course of interviews after the accident that he had considerable difficulty in conducting a conversation in English. However, he was able to cope with all the routine and familiar radio transmissions which had been made up to this point in the flight. The Talkdown Controller appeared to have

instinctively understood the difficulty experienced by the Spanish pilot in understanding the first transmission. His rephrasing in both style and speed is more consistent with common civilian practice: 'One Two Three, you are identified by Northolt Talkdown, your distance eight and a half miles, read back QNH set.'

No further difficulty in radio communication was noted until MAQ123 was 3.5nm from touchdown, at which point the Talkdown Controller said, 'Three and a half miles, check gear, acknowledge'. Once again the speed and style of this transmission is common practice with military controllers. However, the request to 'check gear, acknowledge' is subtly different from standard civilian practice where the pilots are simply reminded to check their undercarriage at this stage but not to 'acknowledge'. The first officer did not understand this transmission but, before he was able to say so, the controller made a second transmission, 'One Two Three, confirm gear down.' Once again, the first officer's knowledge of English could not match the speed of this transmission and he was unable to reply within a reasonable time. The controller then said, 'One Two Three, confirm undercarriage is down'. At this point the commander, who had been trying to explain to the first officer what was requested, took over the radio. However, in the process the commander became distracted from his primary task of flying the approach and the aircraft was seen to deviate above the glide path.

Both pilots were interviewed after the accident. Since a PAR was not commonly available at civilian airfields, the aircraft commander had decided to obtain as much training value from the approach as possible and fly the

Above:
Learjet 25 similar to that which ran off the end of the runway at RAF Northolt in August 1996. *P. R. March*

Learjet to the published minima. He therefore remained on instruments until decision height, when he requested landing flap from the first officer.

The commander stated that he did not appreciate at first that his point of touchdown was considerably further down the runway than was desirable. Once this became clear, he thought about going round again but decided that there was insufficient runway length remaining in which to carry out the manœuvre safely. The first officer said that he was concentrating on the taxi chart in order to guide the commander after landing, and he did not notice that the spoilers had not been selected.

In her statement, the Learjet's passenger stated that there was some disagreement between the pilots at a late stage in the approach and that the commander forcibly removed the first officer's hand from the throttle levers. In the investigators' interviews with the crew, both pilots were adamant that no such disagreement took place. The first officer attributed his own lack of intervention to a complete confidence in the commander's flying ability.

Whatever the personal chemistry on the flightdeck, inter-crew communication was effectively broken when the commander took over radio communications prior to landing. Although the first officer appreciated that the aircraft was high on the approach, he did not ensure that the commander took corrective action. Instead, the first officer focused his attention on what he considered would be his next task–guiding the commander during aircraft taxying after the completion of the landing.

As the aircraft went high at a late stage in the approach, there was little opportunity to reduce the excess speed. Consequently, the Learjet touched down, travelling fast, at a point on the runway with approximately 3,125ft of landing distance left. Notwithstanding the short distance remaining and the speed of the aircraft, it would still have been possible to stop the aircraft on the runway had the spoilers been deployed on touchdown. As they were not deployed, it was not possible to stop in the distance available.

Learjet Inc. does not make any recommendation in the Flight Manual for the ideal time for the selection of full landing flaps . It was standard MAC practice to select full landing flaps late in the approach when a transition to visual flight had been achieved, and the handling pilot had confirmed that the landing was assured. Such a practice has several drawbacks, as the aircraft is flown on the approach in a relatively low-drag configuration such that, on a steeper glide path than normal, speed control is more difficult on a high-performance, low-drag aircraft. In other words, should the Learjet become high on the approach, any attempt to correct the situation by steepening the approach path will result in a consequent speed increase unless corrective action is taken immediately.

The requirement for the first officer to select the full flap position at a late stage might distract his attention from the details of the flight path, reducing his ability to advise the handling pilot on the best course of action to take should corrective action be required. In addition, by selecting flap at a late stage there is always the possibility that the two pilots will both have their hands on the centre console at the same time. Apart from the likelihood of either pilot inadvertently interfering with the actions of the other as a result, to an uninformed observer this may well be interpreted as a conflict between the two pilots, which was the impression formed by the passenger. This perception is even more understandable considering that the passenger did not understand Spanish.

The aircraft commander remained convinced after the accident that he had selected spoilers after touchdown. However, the physical evidence confirmed that this was not the case and he later admitted that he could not actually remember making the selection. The first officer had no recollection whatsoever of the selection or otherwise, as his concentration at this stage in the flight was with the direction in which the aircraft would have to taxi at the completion of the landing run.

*   *   *   *   *

Over the previous year, civilian traffic accounted for 42% of RAF Northolt's fixed-wing traffic. Northolt was so popular among the VIP and charter jet set that the Ministry of Defence limited civilian airfield movements to 7,000 per year. As the principal user of Northolt airfield in terms of numbers of aircraft movements was civilian fixed-wing traffic, the AAIB report thought it highly desirable that standard international civil ATC procedures should be used when controlling civilian air traffic at RAF airfields. Yet civil air traffic was hardly a novelty at the military base and everyone seemed to have coped up to then. Once again, the accident investigation system came up against the flight safety dilemma between 'something must be done' and spending precious money to prevent a one-off human error. Of more practical value was the suggestion that if an ILS/DME system was introduced at Northolt, it would afford crews greater accuracy in assessing their approach and would not require any greater fluency in the English language than that which would normally be required at an international airport.

Had there been a gravel arrester bed at the end of Runway 25 at Northolt, it is undeniable that the Learjet would have stopped more rapidly and would not have reached the A40 road. A review was undertaken of those RAF airfields where a major road is adjacent to the runway end, and RAF Northolt was singled out as having the most serious problem. There are plans to install arrester beds in the overruns of both Runways 07 and 25.

Overall, the flight on 13 August was typical of those carried out by the Spanish charter operator. The planned destination of RAF Northolt was not familiar to either pilot, but that is not rare in charter operations. By virtue of its status as an RAF airfield, the ATC phraseology in use did not accord strictly with the international civil aviation standard, but both pilots had been brought up in the Spanish Air Force, which followed US Air Force-speak.

If the commander had abandoned his intention to remain on instruments until his decision height, he may well have appreciated his position as being both high and fast on the approach and carried out the missed approach procedure. As it was, having not looked up from the instruments until the decision height and having to request landing flap at this late stage, he became sufficiently overloaded that he neglected to select the spoilers after touchdown.

The crucial flight safety point is that, by the time Learjet EC-CKR reached the Northolt runway, the crew had ceased to operate as a team. The commander's aircraft handling was no longer being monitored by the first officer, who was concentrating on the route to the parking area. In normal circumstances this would not have mattered, but the combination of high speed and late touchdown increased the flying pilot's workload to the extent that he neglected to deploy the spoilers on touchdown and this, unnoticed by the first officer, resulted in the overrun. Once the commander had appreciated that the landing was well into the runway and excessively fast, he decided that a go-around was not feasible. A brake parachute was available but not deployed. Perhaps the flying pilot was too overloaded to think about the braking parachute, but, for it to have been effective, it needed to be deployed early in the landing roll.

Although the first officer had turned down the opportunity of a command to avoid the responsibility that such a position incurred, the charter company appeared to have indulgently accepted this situation without regard to the wider implications of his performance as a crewmember. The lack of involvement by the first officer in the operation of the aircraft during the last stage of the flight demonstrated a fundamental ignorance of the principles of Crew Resource Management (CRM) training. CRM, which is geared to helping individuals work together as a team when under pressure, has become the vogue in recent years, and the AAIB report on the Northolt accident recommended that the Spanish Civil Aviation Authority should begin to implement the planned requirements for CRM training as soon as possible. This was excellent advice, if rather unworldly. Learjet EC-CKR was the only aircraft on the books of MAC Aviation when it was founded in 1984. Although the company would grow to five aircraft, a downturn in the charter business meant that by 1996 MAC Aviation was operating only two Learjets, with five pilots, on a mixture of company flights and *ad hoc* charters. The idea that such a small company could afford the luxury of CRM training shows the extent to which some in the aviation authorities need to get out more often. All the courses in the world can never beat a recruitment system that insists on team players, or aircraft commanders who encourage their crew to sing out if they notice anything amiss. Whatever your aircrew position on the flightdeck, you are never along just for the ride.

12

Afterthought

US Navy fighter-bombers dropped 25% of the munitions deposited on Afghanistan during the war against the al-Qaida organisation and the Taliban regime after the hijackings of 11 September. This relatively small air effort (compared with that expended by the US Air Force) came about because targets in landlocked Afghanistan were at the limit of carrier-based fighter-bombers' reach. It took many in-flight refuellings to sustain an F/A-18 Hornet on its round trip to and from the 4½ acres of US real estate floating in the Arabian Sea, and these replenishments did not always go to plan, The following report was written by an anonymous US Navy pilot, to whom all due acknowledgement is made.

'Thought y'all might get a kick out of a recent experience of mine. In case anyone asks, flying around in an F/A-18 without a canopy is bad for the skin. Twenty thousand feet over Afghanistan in an open-air McDonnell Douglas Cabriolet is just a bad, bad place. Air's real dry up there, causes the skin to dry out. That and the wind chill of course.

'01.30hrs launch. Fifth and final planned tanker rendezvous, FL250, 280-285kt with an hour and a half to recovery. Sun wasn't up at 06.00hrs but it was bright enough. My goggles and goggle bracket were both stowed. The KC-10 tanker had finished replenishing itself a half hour earlier and we were immediately afterwards. I was the fifth guy to tank.

'The boom operator called "clear" before I tanked. Tanking appeared normal to me and the air was smooth. Then a length of hose separated. I pulled the power back and picked up the nose. After some wailing and flailing, the KC-10 and I disconnected but I still had the basket and seven feet of hose with me. The hose had a 10lb fitting on it which then proceeded to beat the living **** out of my airplane. "This is gonna be bad, this is gonna be real bad," I thought. I was right. After 20 sufficiently violent whacks the canopy gave up the ghost. I never thought about what a shattering canopy would sound like. Up until then of course.

'I figured since it's made of plastic it shouldn't sound like glass. Wrong! It sounded just like when you fly through a plate glass window. Of course all the glass flew out and cockpit pressure went from eight grand to ambient in about a heartbeat. Which was a pretty small unit of time right then.

'Don't know exactly where the KC-10 went. Last I saw him he was turning for the southwest, spewing gas in the air and words over the radio. *Bossman* had no time for little 'ol me. One of his Air Force brethren was experiencing discomfort. Had to yell at him to get his attention to tell him I was no longer there.

'At first (before I put the top down) I thought I could make it home. Okay, it's 650nm away, I can make it if I go pretty slow and kinda low. And that hunk o' **** on my nose can't be doing much for my gas mileage. This should warrant a ready deck.

'Descended about 3,000ft and decelerated to about 260kt by the time the canopy blew. Then the glass shattered. Okay, Jacobabad it is.[Jacobabad was one of three Pakistani airfields used by the coalition for transport and reconnaissance missions.] Went down to about 19,000ft and put out the speedbrake. Fitting was still beating up the jet while passing through 240kt. At about 230kt the beatings stopped and I started down, maintaining airspeed. Flight controls and engines appeared fine. Ball was a little out of centre but that was it. Didn't have to turn to put Jacobabad on the nose. It was straight ahead.

'Nav system told me it was 260nm away. My body told me it was pretty damn' cold up there. The hose remains were still trying to get at my head so I started descending and decelerating (opposing states, so I'm not sure I did either one that efficiently). Levelled off at 12,000ft. I stopped getting beat up, the fitting just hung in the slipstream by my canopy bow at 230kt. So there I was . . . Eight thousand feet above Afghanistan at 230kt.

'"You know, if a guy really wanted to get shot by a man portable surface-to-air missile he'd fly a profile a lot like what I'm doing right now." Oh well. It's at times like this when you just make a decision and go with it. If you pull it off then it was ". . . outstanding airmanship and in keeping with the highest tradition of the United States Naval Service". If you don't pull it off, maybe they'll name a safety award or the new Base Gym after you.

'My wingman was still with me through all this. Because of some late tankers and shuffling to get guys that were using our tanker to go further north, he only had 10K in gas so he definitely wasn't gonna make it back. Well, not definitely, he could still tank after all. But because of how I had to sit in the cockpit to minimise the windblast I needed him to watch over me.

'Seat lowered, visor down, cockpit heat up full and hunched over staring at one of the TV screens in the cockpit, it's weird the thoughts that come to you during times like this. "You know, sitting this close to the screen is bad for your eyes." Had to snicker over that one. I could look right and left and see the Afghan and then the Pakistan scenery slowly drifting by, too slowly. On the descent the airplane's computer was displaying how long it would take me to get to the divert given my decelerating airspeed. "Okay, twenty minutes not bad. I can do that no pro- . . . oh thirty minutes now. Okay, piece of cake . . . Forty!? ****." Settled out at forty-eight. In the end I didn't really look outside much. Just peeked over the dashboard every couple of minutes to make sure the velocity vector was on top of the upcoming ridgelines. This part of the world is not pretty, by the way.

'Once everyone realised the seriousness of the situation they started to talk to me. The AWACS [Airborne Warning & Control System] aircraft switched me over to the E-2 in charge of the south. They started relaying stuff I needed to tell the

boat, like the parts the jet would need in order to make a flight back out again. "The boat wants to know how badly the canopy is cracked." I couldn't believe that one. I thought he would have heard all the wind in the cockpit and known. "It's not cracked, it's gone. I'm flying a convertible." Apparently that line made it through all the nets loud and clear. The next day I was talking with the Combat Search & Rescue guys in Jacobabad and they said they got spun up when they read that on chat. (It's all real time chat nowadays.) What did not get through was the driver of the convertible. I know the E-2 guy knew who I was (the conversation by the end had degenerated to callsigns Gretzky and Duck.– not professional but somewhat comforting) but somehow the ship was waiting for me to return at 09.00 vice my wingman. All this technology.

'As far as the cockpit was concerned there were two different and distinct regions. From my knees down I was toasty and warm — "this little piggy" was getting sweaty, in fact. The chilly zone came above that. The wind was swirling around pretty good and I was trying to grab all the paper and shove it into my helmet bag. Only lost one bit of classified stuff. Not too bad, all things considered. After 20 minutes I started getting the

Above:
While a US Navy F/A-18 Hornet refuels from a US Air Force KC-10A tanker, an F-14A Tomcat waits its turn along with two more F/A-18s.
US Navy/Cdr. Brian G. Gawne

shakes; after 30 they were fully developed. I tried to stuff my whole body down by the rudder pedals, with limited success. Kept my hands warm though — thank God for autopilot.

'About this time my wingman came up and said, "Hey can you reach out and grab that thing, pull it in?" I looked over at him (not that he could see me) with a look of shock. Stick my arm out into that wind, get my arm blasted back and thrashed on the glass shards sticking up everywhere? "Have you lost your mind?!" "Oh yeah, guess it's kinda windy. Sorry." Like I said, it's strange the thoughts you have sometime.

'My wingman and I talked about the airfield — frequencies, layout, the fact that the locals like to shoot at planes landing there. You know, just normal airport talk. We talked about landing on a runway, something neither of us had done for three months. And we dumped fuel to lighten the load. We both were carrying 2,000lb of unexpended ordnance, so the Air Force guys were gonna love us. Lastly we dropped the landing gear in close formation and compared airspeed and AOA (Angle of Attack) to make sure the KC-10 hadn't damaged my AOA and airspeed probes as well. I had him land first because I thought the hose might drag on the ground and get rolled up on by the nosewheel. After that who knew what would happen?

'Approach was initiated from 5,000ft agl when the threshold was 10° down. Started to slow the descent at about 500ft. Landed on speed at the nine board. Don't remember seeing a Visual Approach Slope Indicator or anything. Airfield diagram on approach plate doesn't show any landing aids.

'The plane flew fine with all that junk on it. Just had to use the rudder pedals, which is kind of an emergency procedure for a Hornet pilot. When I slowed I got the "sunroof effect" pretty bad. You know, when you're zorching down the road and you open the sunroof but leave all the other windows up? That vibration you get until you crack another window? Well I got kind of an advanced case of that during my Space Shuttle descent to final. We both rolled out fine. Well, maybe not fine. We had to use all 10,000ft and both had smoking brakes. (Our brakes hadn't been used like that in awhile. On the boat the arrester wire brings you to a gentle stop without them, of course.) The emergency crews were waiting for us. And they were pointing and gawking as would be appropriate for a situation such as this. Couple of natives looked on in a disinterested manner.

'Of course I had to do a flight physical after all this. Had to make sure I wasn't on drugs before I launched on my six-hour mission into Afghanistan. The facilities in Jacobabad ain't that bad. I'm here to tell you we are number one in tent technology. Our tents kick ass. They got AC and everything. Since it's an Air Force base they got all the best entertainment. Drew Carry and Joan Jett had been there already. Shania Twain was supposedly coming too (broke my heart–if only I'd had better timing . . .). And, of course, the Toga Party on Saturday. Can't forget that. Yeah, it's kinda like the boat. Except for the booze and the Toga Parties. Other than that it's just like the boat. Other random observations: Air Force got all the good buildings. Marines are on the outskirts, again. The boys from the 101st Airborne are spoiling for a fight. Hate coming in behind the Marines all the time. Dust over everything. Lots of people there that don't look like they are in the normal military.

'The maintainers showed up about four hours after I did. After the appropriate amount of gawking they got to work and fixed it well enough for the RTB (Return to Base) in under four hours. Nice job all around. By the end the basket and hose were removed, the canopy had been replaced and the airframe repaired with 300-mile-an-hour tape. The Air Force settled all the maintainers into two spare tents and they had a grand ol' time.

'The next morning I took off low and fast at sunrise. Low and fast was due to the locals and the guns of course, not because it was fun. I checked in and the E-2 said, "It's good to hear your voice again." The RTB was uneventful right up until the end. A PTS shaft died and subsequently one of my hydraulic systems gave up the ghost when I dropped the gear. I got a couple of spurious flight control cautions but didn't really give it much thought as I was working the landing. As I started the approach turn the nose started to wander and I got another caution tone. I lost one aileron, one rudder and half a horizontal stabiliser. I hit the reset button and I think everything cleared. Then I saw the Hydraulic Cautions come up. Hitting the reset button to go suddenly from normal response on short finals was a big mistake. When the aileron failed again I realised I sorta needed to get aboard the first time. "Man, first I miss Shania and now this. This is just not my week." I got it aboard because the Hornet is a fantastic jet. I got a Fair grade for the pass because I'm not very smooth when I'm rattled.

'I pretty much assumed I was in trouble throughout all this. A canopy has got to cost 70 or 80 grand. Depending on how much repairing the windscreen and the airframe were . . . it could cost over 200 grand. Which would mean a Class B mishap. Which would mean I was screwed. Again. Thinking all this and then seeing the CO waiting for me when I landed made my heart sink. But that was not the reason he was there. The decision was made somewhere to make a big deal about this in a good way. Just like that . . . dirtbag to hero. Funny. Turned over bodily fluids to VFA-147 Safety Officer.

'This isn't the first thing that's happened to me out here, you know. We're flying the **** out of these jets and it's starting to show. I had to come back from the box with an engine shut down a week or two before. I'm getting too short for this ****. Oh well, statistically speaking the rest of the cruise should be smooth sailing. What are the odds something like this will happen again?'

All of which goes to prove that whatever advances science and technology bring to aviation, aircrew need to be ready and prepared to deal with the unexpected. The ultimate responsibility of any pilot is to maintain control of his or her aircraft, and automation must never reach the stage where the human is pushed too far out of the loop.

The difference between safety and catastrophe is usually an alert someone who knows where they are and is able to respond flexibly to whatever fate throws at them. Situational awareness is everything in the air, especially when things start to go wrong and inevitable distractions crop up. And that awareness comes from airmanship built up over years of flying experience, which means being used to flying in all weathers and knowing your stuff. Anybody can learn from their own mistakes. The wise pilot learns from the experience of others that flight safety is no accident.

Index

Dates refer to specific accidents